Z - SCORE

How a Statistic Used in Psychology Will Revolutionize Baseball

John G. Cottone, PhD
&
Jason Wirchin

Foreword by David Lennon, National Baseball Writer, *Newsday*

Story Bridge Books
Tucson, AZ

Z-score: How a Statistic Used in Psychology
Will Revolutionize Baseball

Copyright © John G. Cottone, PhD,
Jason Wirchin & David Lennon, 2015

Published by Story Bridge Books
Tucson, AZ

ISBN: 978-0-9903939-3-1
Library of Congress Control Number: 2015934569

Written by
John G. Cottone, PhD
&
Jason Wirchin

Edited by
Laura Markowitz

Foreword by
David Lennon

Cover Design & Artwork by
Rook Designs:
John Tsopanarias, Jr & Lisa A. Cottone, PhD

Copyright © 2015

All rights reserved. Without limiting the rights under copyright reserved above, no part of this book may be reproduced, stored in or introduced into a retrieval system, or transmitted, in any form, or by any means (electronic, mechanical, photocopying, recording, or otherwise), without the prior written permission of the copyright owner.

Illustrations:

Original Cover Design:

Copyright © **Rook Designs**, 2015
John Tsopanarias Jr. & Lisa A. Cottone, PhD

Original Images Used with Permission for Cover Design:

(Vintage Baseball Isolated on White)
Copyright © Shutterstock.com/ 222088543.
Used with permission.

(Antique Scale)
Copyright © Thinkstock.com/149086282.
Used with permission.

(New Baseball Isolated on White)
Copyright © Thinkstock.com/493246341.
Used with permission.

(Syringe)
Copyright © Thinkstock.com/177316751.
Used with permission.

Original Figures Used in Text:

Lisa A. Cottone, PhD
Copyright © 2015

Figure 1.1: The Bell Curve (i.e., The Curve of Normally Distributed Scores).

Figure 1.2: Final Grades for Mr. Hardnose's Class, Plotted on the Bell Curve.

Figure 1.3: Z-scores and Percentiles Plotted on the Bell Curve up to $\pm 5\sigma$.

DEDICATION

I would like to dedicate this book to the memory of Dr. Robert Youth. Dr. Youth was the first to introduce me to the concept of z-scores in the *Statistics for Psychology* class that he taught at Dowling College. He was also an avid baseball fan and attended many of my college baseball games. In many ways, he was like a second father to me. He will be sorely missed but his memory lives on in the lessons he taught about statistics, psychology, baseball and life – particularly those lessons that helped form the basis of this book.

- John G. Cottone, PhD

To my grandfathers, for embodying the spirit of baseball's heyday; Dad, for indoctrinating me into the Flushing Faithful; Ben, for always making sure I know who's boss...and Uncle Marc, for knowing that "Ya Gotta Believe," even in the bottom of the ninth.

- Jason Wirchin

ACKNOWLEDGEMENTS

We would like to acknowledge the following individuals for the invaluable feedback they provided in the realm of statistics consultation:

Jeffrey Nevid, PhD, ABPP
St. John's University

Richard Yi, PhD
University of Maryland

Art Aron, PhD
Stony Brook University

FOREWORD

David Lennon
National Baseball Columnist
Newsday

Every November, the long, thin envelope shows up in my mailbox, postmarked: Cooperstown, New York.

The contents? A puzzle, really. A list of names that contains more questions than answers, with few instructions to follow. It's up to me and more than 500 other members of the Baseball Writers Association of America to make sense of it all, to provide some sort of compass, despite trying to follow a historical path that has been blurred by disorienting opinions and an ever-changing landscape.

Such is the voting process for the Hall of Fame, where the credentials for immortality can be hard to define.

Just like me, you've been part of these endless debates. Over the dinner table at Thanksgiving. Having a drink with buddies after work. Sitting in the Pepsi Porch at Citi Field. Plenty of discussion. Zero resolution.

Well, what if I told you that someone finally had come up with an answer sheet? Ladies and gentlemen ... the z-score.

As impossible as it may seem to compare Dazzy Vance to Pedro Martinez or Babe Ruth to Barry Bonds, there is a now a unit of measure, explained in the subsequent pages, to do precisely that. To give every performance context – according to the specifications of that player's era – and then, of course, allow us to rank them right down to the decimal point. And you never have to pick up a calculator.

In this data-driven age, we can feel a bit overwhelmed by statistics, which are readily available at the tap of our phone's touch screen. Personally, when it comes to baseball, I'm probably somewhere in the middle between the Old Schoolers and the Sabermetric crowd. I can appreciate the intangibles of a players' makeup – intellect, confidence, flair for the big moment – along with the meticulously-detailed numerical analysis that quantify his talents.

But the evolution of baseball, from 1876 to 2015, has created an uneven playing field for such analysis due to adjustments to the equipment, the mound, the schedule – and even significant changes to the players themselves. Because of that, the way we view the game, from decade to decade, must adapt as well.

Early on, we trusted batting average to tell a hitter's story, if only because it was one of the more basic calculations to make, other than simply tallying home runs and RBI. Ted Williams became a legend for batting .406. The .300-plateau remains the benchmark for separating a very good hitter from an immortal one.

But now, with piles of statistics to choose from, it seems that we've actually turned greatness into more of a gray area. A player can be productive on many different levels – with a variety of metrics to illustrate his value – and we're again left debating which number carries the most weight.

For baseball, a sport that treasures its records like no other, we in the media have the difficult task of affixing importance to these numbers. Not necessarily creating the formulas – not me anyway – but telling the story behind them, and where they fit in our perception of the game.

This has become particularly challenging for my generation, which witnessed the explosion of performance-enhancing drugs – specifically steroids – toward the end of the 20th century. Suddenly,

Maris' single-season record of 61 home runs was no longer an unreachable peak, but merely a speed bump as Mark McGwire, Sammy Sosa and soon after Barry Bonds soared above it. Even Hank Aaron's 755-homer barrier would fall, steamrolled over when Bonds got to 762.

That left us – you, me, the BBWAA – wondering what it all meant. Is Bonds the greatest home-run hitter of all-time? Is Aaron? Is Maris or Ruth? And how do we treat this new breed of supermen going forward? Should we disqualify a whole generation of players because of the PED stain, even if many are only fingered by suspicion?

These cases have been heard in the court of public opinion for almost a decade now. But when it comes to voting for the Hall of Fame – and even the annual awards like MVP and Cy Young – there is a pressing need for context. As a baseball columnist, I'm faced with that question every season, and each time that ballot arrives at my doorstep.

That's why the concept behind the z-score made for such a compelling read, and better yet, the results it yielded. As you'll discover, Ruth's 29-homer season in 1919 was possibly the greatest offensive accomplishment in baseball history – as calculated by its z-score of 5.70. Bonds' 73 home runs in 2001? That ranks 18th in Cottone and Wirchin's analyses with a z-score of 3.42, one spot below Jose Bautista's 54-homer season in 2010.

Unfortunately, the z-score can't tell us who was actually taking PEDs and when. That we'll probably never know. But by comparing those comic book-sized players with each other, we do get a clearer picture of the effect PEDs had on the game, just like the reverberations from increasing the diameter of the bat in 1895 or lowering the mound in 1969.

What will spur the next big statistical revolution in baseball? Hard to say. Now that we've apparently cleared the Steroid Era, offense has plummeted, and the buzzword these days is "run-prevention" – or what we used to call playing good defense. Regardless, it's all part of the equation going forward, ready for processing.

So consider yourself lucky to have this book. Next time the argument comes up that Don Mattingly would have been a better choice than Roger Clemens for the 1986 American League MVP, you'll have the right answer.

And that's only the beginning. Everybody has an opinion when it comes to valuing players, but it's more important to have evidence. Now you will.

TABLE OF CONTENTS

INTRODUCTION – 3

CHAPTER 1   WHAT THE f IS A Z-SCORE? – 13

    INTERIM SUMMARY: THE LEAST YOU NEED TO KNOW (ABOUT Z-SCORES) – 38

CHAPTER 2   RUNS, HITS & ERAS: A SUMMARY OF BASEBALL HISTORY BY ERA – 40

CHAPTER 3   PRIMETIME LINEUP: BASEBALL'S Z-SCORE LEADERS – 80

CHAPTER 4   PITCHERS WINNING THE MVP – 110

CHAPTER 5   REVENGE OF THE NERDS: AN HOMAGE TO *MONEYBALL* – 132

CHAPTER 6   Z-SCORES & OTHER SPORTS – 151

CHAPTER 7   HALL OF FAME ELIGIBILITY – 160

CHAPTER 8   FUTURE DIRECTIONS – 168

APPENDIX A   GLOSSARY OF ABBREVIATIONS – 177

APPENDIX B   REFERENCES – 179

APPENDIX C   FORMULAS – 183

APPENDIX D   COMPUTATION EXAMPLES – 184

INDEX – 187

ABOUT THE AUTHORS – 195

"Baseball is 90% mental. The other half is physical."

- Yogi Berra,
New York Yankees

INTRODUCTION

JOHN G. COTTONE, PHD

"Allow myself to introduce… myself."

- Austin Powers
Austin Powers: International Man of Mystery

It was less than a month after September 11, 2001, when, in an effort to distract myself from my anger, fear and sadness, I initially conceived of this book. I had just started the 3rd year of my psychology doctoral program at St. John's University, and I was reflecting on the data analyses I was planning for one of my research projects. The date was October 5, 2001, and on ESPN I watched as Barry Bonds hit his 71st and 72nd home runs of the season, breaking the record set by Mark McGwire just three years prior. McGwire hit 70 home runs that year, as he and Sammy Sosa raced to erase Roger Maris' record of 61 home runs, set in 1961. Maris' record had stood for nearly 40 years, and now, as I watched Barry Bonds circling the bases after taking a pitch deep from Chan Ho Park, the major league home run record had been broken for the third time[1] in just three years.

When Sosa and McGwire broke Roger Maris' record, there were whispers about them using steroids. However, by the time Bonds eclipsed McGwire's record, most baseball fans were convinced that he was using steroids, even though there was no definitive proof at the time. Part of the reason for suspicion was not just because Barry Bonds was about to accomplish this astonishing feat at the age of 37 – something he hadn't come close to doing during the supposed "prime years" of his career a decade earlier – but because home run totals across the league seemed to have been artificially inflated for the past 15 years. I remember growing up in the mid-1980s when Dale Murphy led the National League in home runs for two straight years – 1984 and 1985 – with 36 and 37 home runs, respectively. Yet, had Dale Murphy hit his 36 home runs in 2001, when Bonds set *his* record, Murphy would have finished 13th in the National League and 22nd in the majors.

Appraisals like these beg one to ask questions like: How do Barry Bonds' 73 home runs in 2001 compare with Roger Maris' 61

---

[1] Including Sammy Sosa's breaking Roger Maris' record during the 1998 season and his leapfrogging of Mark McGwire, as he briefly held the home run record on 9/25/98 with 66 home runs – a total with which he would finish the season.

home runs in 1961 (or Babe Ruth's previous record of 60 home runs, set in 1927), given that Bonds set his mark in an era of rampant steroid use in baseball; when ballparks, on average were much smaller than they had been in previous eras; and when baseballs were believed to be wound tighter (i.e., more "juiced") than ever before? I, like the rest of America, asked questions like these repeatedly during Barry Bonds' historic home run chase, and what I discovered in my statistics classes, during the fall of 2001, is that the use of z-scores could be a tremendous aid in answering such questions.

For those who have not had the "privilege" to struggle through a dry statistics course, and thus, may not have heard of z-scores, I'll explain: z-scores are a universal form of measurement that allow for more sophisticated comparisons than what is possible using only raw data. As will be explained in more depth in Chapter 1, rather than providing a direct and absolute form of measurement, each z-score describes an individual's performance *relative* to the performance of others in the same pool (e.g., in the case of Barry Bonds' 73 home runs in 2001, his z-score would describe his home run performance relative to the home run totals of all other National League hitters during the 2001 season).

One important reason for using z-scores to evaluate baseball statistics is that they allow you to make sophisticated comparisons within a given statistical category across eras of baseball history. For example, in comparing Barry Bonds' 73 home runs in 2001 with Roger Maris' total of 61 in 1961, you can determine which accomplishment was more statistically rare, as the z-score for both Bonds' and Maris' feats accounts for the number of home runs their peers hit during those respective seasons. Additionally, using z-scores allows one to compare data across different statistical categories that use different metrics: For example, you can directly compare Babe Ruth's 60 home runs in 1927 to Pedro Martinez's 1.74 ERA in 2000, to Wayne Gretzky's 92 goals in the 1981-82 hockey season, to Manute Bol's towering height of 7'7", to Albert Einstein's supposed IQ of 150, *to the price of tea in China in 1940...*

Pound for Pound

Within each sport, as well as between sports, z-scores can help us to determine the best performers, *pound for pound*. The term *"pound for pound"* is often used in boxing as means of describing a fighter's skill level, regardless of his weight class. Essentially, when a boxing analyst suggests that a given fighter in the Welterweight Division (140 – 147 pounds) is, pound for pound, a better boxer than one in the Heavyweight Division (over 200 pounds), what the analyst means is that if the two fighters were the same weight, the welterweight fighter would likely win. For instance, when I was in college, struggling on the pitching mound due to my allergies to the strike zone, welterweight fighter Floyd Mayweather, Jr. was considered to be the best pound for pound boxer in the sport. Considering that Mayweather, Jr.'s career spanned an era that was also occupied by heavyweights like Mike Tyson, Evander Holyfield, and George Foreman, this distinction is particularly impressive. In essence, boxing analysts believed that if Mayweather, Jr. (who weighed around 147 pounds for much of his career) was the same weight as Mike Tyson, Evander Holyfield, and George Foreman, respectively (each of whom weighed above 200 pounds during their boxing careers), he would stand a good chance of beating each fighter.

Of course, in combat sports like boxing and wrestling, a fighter's weight plays a very important role in the application of his skill, which is why fighters are divided into weight classes in the first place. Hence, it is impossible to fairly assess a fighter's skill taking weight out of the equation. As such, efforts to this point to compare fighters across weight categories have had to rely on *subjective*, expert opinion. However, for those who are interested in a more *objective*, empirical approach to comparing performance *pound for pound* – whether it be in boxing, or any other sport, within the present or across eras in history – we believe that using z-scores can be a helpful tool in this endeavor. Though the use of z-scores in this regard is not a perfect solution, we believe that z-scores offer a more objective assessment in such comparisons than sole reliance on expert opinion.

The Forefathers

It is written in the Ecclesiastes in the Bible (1:9): "…what has been done, will be done again, and there is nothing new under the sun!" This book does not represent the first attempt to use z-scores to evaluate sports performance. If you do a Google search for "z-scores and sports statistics," you will find numerous blogs that discuss the topic, and a few that even offer some data. You may also come across a book written by biostatistician, Michael Schell, published in 2005 by Princeton University Press, entitled: *Baseball's All-Time Best Sluggers: Adjusted Performance from Strikeouts to Home Runs*. When he is not busy writing books about baseball statistics, Michael Schell is performing data analyses to help develop treatments for cancer, and at the time of publication of this book, Schell was the Director of the Biostatistics Core Facility in the Lineberger Comprehensive Cancer Center in North Carolina. His contributions to cancer research not only dwarf in importance the findings of this book, but the athletic accomplishments of the athletes we *both* analyzed.

In *Baseball's All-Time Best Sluggers*, Michael Schell takes a slightly different approach than we do. First, Schell does not report z-scores as a stand-alone statistic in his book: rather, he uses z-scores to compute his own series of unique, adjusted batting statistics. Second, Schell's book only examines baseball and focuses almost exclusively on batting. *Z-score* differs from *Baseball's All-Time Best Sluggers* in that we report z-scores as a stand-alone statistic (not as a component in an algorithm) because we feel that these data themselves provide invaluable information. Also different is that *Z-score* reports analyses for batting *and* pitching statistics, and we also provide analyses in Chapter 6 for the other three major American team sports: football, basketball and hockey. Though our books are different, after reading *Baseball's All-Time Best Sluggers*, I offer high praise to Michael Schell for what he was able to accomplish and I highly recommend *Baseball's All-Time Best Sluggers* to those of you who are interested in sabermetrics, fantasy baseball and a more sophisticated way to analyze baseball statistics.

At this point, we would be remiss if we did not mention some of the other giants in the field of sports statistical research. In the preceding paragraph, the term "sabermetrics" was used, and it is a testament to how far things have progressed that most sports fans understand what the term means without further explanation. However, for those of you who may be unfamiliar with sabermetrics, a brief explanation is in order. Sabermetrics is a term derived from the acronym SABR, which stands for the Society for American Baseball Research, which was founded in 1971 by L. Robert Davids. For the past 30 years, however, Bill James, has been the poster child for SABR, and he is the one credited with coining the term "sabermetrics" to refer to the advanced statistical analyses pioneered by SABR, or in the spirit of SABR's mission to develop more objective tools to analyze baseball.

Though not the father of sabermetrics, Bill James might well be considered *the Godfather* of sabermetrics. Since 2003, Bill James has been a senior advisor to the management of the Boston Red Sox and he was one of the innovators, who, along with Theo Epstein, helped the Red Sox end the Curse of the Bambino in 2004. However, despite his influence with the Red Sox, Bill James might be best known for his influence on Billy Beane, the general manager of the Oakland A's (since the end of the 1997 season). Beane's use of sabermetrics to counter the financial advantage that rich teams, like the Yankees and the Red Sox, have over poor teams, like the A's, spawned the bestselling book by Michael Lewis entitled *Moneyball*, and a major motion picture by the same name.

In Chapter 5, we discuss *Moneyball*, and how the SABR/James/Beane revolution changed the way we all look at baseball by changing the markers of success for individual performance. In short, James, Beane and the other sabermetricians huddled in the dark corners of baseball's underground reasoned that if the object of baseball is to win games, and the way to win games is for a team to score more runs than its opponent, teams should focus on player statistics that best predict how to *score runs* while batting, and *prevent runs* from scoring runs while in the field. It sounds quite obvious when stated so simply, however, the SABR/James/Beane innovation was in determining that some of the traditional measures of both offensive performance (e.g., batting average and RBI) and pitching performance (e.g., ERA and wins) were less predictive of team run production and team run prevention than some of the new statistics that they were developing to evaluate performance, like OPS (i.e., On-Base Percentage-plus Slugging Percentage), wOBA (weighted On-Base-Average, which adjusts for each type of offensive event differently) and Runs Produced (i.e., Runs Scored, plus RBI, minus Home Runs); and for pitchers, WHIP (Walks, plus Hits, divided by Innings Pitched).

The analyses and statistics developed by Bill James and his colleagues at SABR are sophisticated and revolutionary. It would be naïve and arrogant for us to suggest that with *Z-score* we are taking the ball further down the field when in fact we are simply taking the ball in another direction. We believe that, on mathematical grounds, the statistics developed by SABR, Bill James, and Michael Schell are valid and sound. However, these sabermetric statistics have two limitations: a) they are baseball-specific and thus cannot be used to compare players across sports, or evaluate performance in any other facet of life besides baseball; and b) they are very sophisticated, with each one requiring an in-depth explanation for understanding. On the contrary, we believe that z-scores offer a way to not only compare baseball performance to that within all other sports, but they can also be used to evaluate performance in many other areas of life, like SAT scores, median household income and demographic data. And, furthermore, once you can grasp the concept of what a z-score is (discussed in depth in Chapter 1), you don't need to learn the components of any other sophisticated statistic because z-scores can be used to apply data adjustments to all of the traditional statistics used in sports, as well as any of the newer sabermetric statistics being developed each year, and any other type of statistic that may one day be used to evaluate performance in any aspect of life.

Rounding Third and Heading for Home

In the years that have followed Barry Bonds' record-breaking season, the baseball world has been locked in an endless debate about how to compare his achievement with home run records of yore, with some proposing Bonds' record have an asterisk next to it. However, as we continue to learn more about the culture of steroid use in baseball, it is becoming increasingly clear that Bonds' behavior was less the exception and more the rule. In fact, it is likely that a significant percentage of Bonds' record-setting home runs in 2001 were hit against pitchers who were also using steroids.

As will be discussed in subsequent chapters, steroid use is only the latest in a series of significant influences on baseball performance making it difficult to compare the raw statistics of players across eras of baseball history. I was reminded of this sometime in 2011 during a conversation with my longtime friend and baseball aficionado, Jason Wirchin. Our discussion about the changes in baseball across eras was so captivating that I invited him to collaborate with me on *Z-score*, recognizing that his knowledge of baseball history could add life to the dry statistics of the book and make the narrative more colorful. I think most readers will agree that *Z-score* is far more entertaining than it would have been without Jason's witty, informative contributions, which are presented throughout the book but are especially abundant in Chapter 2.

It is our hope that the use of z-scores can put the statistics for players of ALL eras in baseball history on *an even playing field* and thus help fans, historians and even Hall of Fame voters to make fairer evaluations when comparing the great players of our national pastime. We also believe that using z-scores to assess baseball performance could also help scouts, general managers and player agents better evaluate talent across all levels of the sport, thus revolutionizing the business side of baseball.

In the chapters that follow, we will first take you back to school (don't worry, no pop quizzes), to teach you the basics of what a z-score actually is. Next, we will use z-scores to reveal some extraordinary baseball achievements that have been overlooked until now (like Gavvy Cravath's 24 home runs in 1915) and also put some of baseball's most overrated achievements into proper perspective. We will then take a brief detour off Baseball Boulevard to examine z-scores for some of the outstanding achievements in the other three major American team sports and finish by showing some other intriguing applications for our *star* statistic.

So stick with us and get ready for a journey filled with surprises. It's going to be a fun ride and we promise to give you an 'A' just for showing up to class.

Chapter 1

What the f is a Z-score?

"Billy [Beane] was forever telling Paul [DePodesta] that when you try to explain probability theory to baseball guys, you just end up confusing them."

- Michael Lewis, Author,
Moneyball

Note: While an appreciation of the mathematics presented in this chapter is helpful for a deeper understanding of z-scores, for those who may not be interested, we have provided a simplified summary at the end of this chapter highlighting "the least you need to know" about z-scores.

Mr. Hardnose and Mr. Goodgrade

Imagine for a moment that you are back in high school, competing with your rival, Jack Silverspoon, for valedictorian. It's the last week of school and the final class rankings have just been posted. Your heart drops as you check the list and notice that you've been ranked second in the class, just behind Jack. Jack stole your girlfriend Heather at the Senior Prom, showing up in a new convertible that his parents got him, and now he just beat you out for valedictorian.

Dejected, you look for answers. In your search you discover that what nudged Jack ahead of you was that he finished chemistry with a final grade of **99**, while you only got an **83**. You think back to chemistry and how difficult it was, and how it seemed that your teacher, Mr. Hardnose, believed that for every 'A' he gave out a year would be taken off his life. How could Jack have gotten a **99** in chemistry while you only got an **83**? Everyone knows you are smarter than Jack, you work harder, and your mother is a freakin' chemistry professor! Despite your disdain for Jack, you know he's an honest guy, so it's not likely that he cheated.

You reflect a little longer and then it hits you: Jack didn't have to take chemistry with Mr. Hardnose, he had Mr. Goodgrade – a guy who decided to *mail it in* that year because he was about to retire to Arizona. He couldn't be bothered with actually grading test papers, so he just handed out 'A's like the military hands out uniforms. In fact, Mr. Goodgrade gave almost everyone in his class an 'A+' just for showing up (except, of course, for the kids who taped a "kick me" sign to his back on the first day of school). How unfair: you had to take chemistry with Mr. Hardnose, where the average grade in the class was a **68**, while Jack got to slide by with Mr. Goodgrade, where the average grade was a **93**.

You think to yourself that if there was justice in the world, your school would have taken into account that your chemistry class was much harder than Jack's and weighed your scores accordingly.

This would have validated the fact that the **83** you got in Mr. Hardnose's class, which was the second highest grade in the class, was actually more impressive than the **99** that Jack got in Mr. Goodgrade's class. But how can you prove this? What evidence do you have? Enter the z-score!

But what the *f* is a z-score?

Z-scores are a type of standardized statistic that describes an individual's performance, *relative* to the performance of others in the same group or **distribution** of interest. When scores from any distribution are converted to z-scores, this process automatically controls for factors that might influence the whole distribution (like Mr. Goodgrade deciding to give almost everyone in the class an 'A+,' just for showing up) and distort these comparisons. This is because inherent within every z-score is information about two important components of the distribution from which it came: the **mean**,[2] or average score of the distribution (e.g., the average grade in Mr. Goodgrade's chemistry class) and the **standard deviation** (SD) of the distribution – a statistic that describes how far apart the scores of a distribution are from the mean. As illustrated below, z-scores are computed by subtracting the mean of a distribution from a particular score, and then dividing by the standard deviation of the data set.[3]

$$\textbf{Z-score} = \frac{(Score\ X - Mean)}{Standard\ Deviation}$$

---

[2] The mean is calculated by adding all scores in a distribution and then dividing by the number of scores in the distribution (*N*). The formula is:

$$Mean = \frac{(Sum\ of\ All\ Scores)}{N}$$

[3] Lockhart, R.S. (1998). *Introduction to Statistics and Data Analysis for the Behavioral Sciences*. W.H. Freeman, New York.

Using the equation above, the computation of a z-score starts with the specific score you are looking to convert – for example, the final grade you received in Mr. Hardnose's class, which was an **83**. From that you subtract the mean (i.e., the class average), which you recall from Mr. Hardnose's class was **68**. Now, this is the easy part: you subtract **68** from **83** and get **15** in the numerator of the fraction.

$$\text{Z-score} = \frac{83 - 68}{\textit{Standard Deviation}}$$

$$\text{Z-score} = \frac{15}{\textit{Standard Deviation}}$$

The denominator, however, which involves the computation of the standard deviation, is a bit more challenging to compute and explain. As noted above, the standard deviation of a distribution refers to how far apart scores are in a distribution relative to the mean. The formula for the standard deviation is a bit complicated, and not necessary to comprehend to understand the meaning of the z-scores in this book; however, we have provided it below for those who may be interested.

$$^{4}\textbf{Standard Deviation (SD)} = \sqrt{\frac{\Sigma\,(\textit{Score X} - \textit{Mean})^2}{N}}$$

---

[4]Presented here is the formula to compute the standard deviation of an entire population, rather than a specific sample. We presented this version of the formula here to help readers more easily calculate the standard deviation in the example on the following page. However, for our analyses, we used the formula for the standard deviation that corresponds to the standard deviation of a sample:

$$\sqrt{\frac{\Sigma\,(\textit{Score X} - \textit{Mean})^2}{N-1}}$$

Though a comprehensive discussion of the standard deviation is unnecessary for the purposes of this book, one thing about the standard deviation that is potentially helpful to know to better understand z-scores is that two distributions can have the same mean, but very different standard deviations. For example, imagine that two classes with 10 students – Class A and Class B – take the same spelling test, as illustrated in Table 1.1 below. In Class A, five students get a score of 100 on the test, while the other five students get a score of 20. In Class B, however, five students get a score of 70, while the other five students get a score of 50. Interestingly, what you will notice in Table 1.1 is that while the average test score in both classes is 60, the standard deviation for the two classes is very different: for Class A it is 40, while for Class B it is 10!

TABLE 1.1: SPELLING TEST SCORES FOR CLASS A & CLASS B

CLASS A	CLASS B
100	70
100	70
100	70
100	70
100	70
20	50
20	50
20	50
20	50
20	50
Mean (Class A) = 60	*Mean (Class B) = 60*
Standard Deviation (Class A) = 40	*Standard Deviation (Class B) = 10*

NOTE: The actual computation of the standard deviation for Class A and Class B is presented in Appendix D.

As shown in Table 1.1 (and Appendix D), the scores in Class A are more widely dispersed (ranging from 20 to 100) than the scores in Class B (which merely range from 50 to 70), which is why the standard deviation for Class A is much greater than the standard deviation for Class B. This type of information, captured by the standard deviation, is extremely important to statisticians and researchers because it provides the context necessary to better understand the data they interpret. As it relates to z-scores, since the formula used to compute them is a fraction with the standard deviation as the denominator, the larger the standard deviation, the smaller the z-score (assuming all else is equal), and vice versa.

Getting back to you and Jack… now that we've discussed all of the components comprising a z-score, we can demonstrate how your **83** in Mr. Hardnose's class was actually more impressive than Jack's **99** in Mr. Goodgrade's class.

As shown in the computations of Tables 1.2 and 1.3 on the following page, while the **83** you got in Mr. Hardnose's class yielded you a solid z-score of **<u>1.26</u>**, Jack's final grade of **99** yielded him a paltry z-score of just **<u>0.49</u>**! Why? Mostly because the z-score for Jack's final grade (**99**), took into account that the average grade in Mr. Goodgrade's class was a **93**, while the z-score for your **83** took into account that the average grade in Mr. Hardnose's class was a **68**. Hence, *your* grade was much better relative to *your* classmates than was Jack's relative to *his* classmates. Your grade was 15 points higher than your class average, while Jack's grade was only 6 points higher than his class average.[5]

---

[5] Note: In this example, the standard deviation (or *spread* of the scores) was roughly the same for the two classes (11.88 vs. 12.17) and as such, had less impact on the z-scores than did the differences in the class means.

TABLE 1.2: FINAL GRADES – MR. HARDNOSE'S CHEM. CLASS

CLASS MEAN = 68
CLASS STANDARD DEVIATION (SD) = 11.88

STUDENT	FINAL GRADE	Z-SCORE = (SCORE X – MEAN)/SD	Z-SCORE
Margaret	95	(95-68)/11.88	2.27
*** YOU ***	83	(83-68)/11.88	1.26
Joe	75	(75-68)/11.88	0.59
Jane	68	(68-68)/11.88	0
Rachel	62	(62-68)/11.88	-0.51
Chris	60	(60-68)/11.88	-0.67
Melissa	60	(60-68)/11.88	-0.67
Doug	60	(60-68)/11.88	-0.67
Randy	59	(59-68)/11.88	-0.76
Alyssa	58	(58-68)/11.88	-0.84

TABLE 1.3: FINAL GRADES – MR. GOODGRADE'S CHEM. CLASS

CLASS MEAN = 93
CLASS STANDARD DEVIATION (SD) = 12.17

STUDENT	FINAL GRADE	Z-SCORE = (SCORE X – MEAN)/SD	Z-SCORE
Julian	100	(100-93)/12.17	0.58
Francesca	100	(100-93)/12.17	0.58
Alison	100	(100-93)/12.17	0.58
Evelyn	100	(100-93)/12.17	0.58
Sarah	100	(100-93)/12.17	0.58
Joshua	100	(100-93)/12.17	0.58
*** JACK ***	99	(99-93)/12.17	0.49
Norm	93	(93-93)/12.17	0
Robbie	69	(69-93)/12.17	-1.98
Barbara	69	(69-93)/12.17	-1.98

Grading On a Curve

After any difficult exam, the most welcome words a student can hear from the teacher is: "This exam is being graded on a curve." Hallelujah! But, what exactly does this mean? Well, when statisticians or teachers talk about "the curve," what they are referring to is the **"bell curve"** or the curve of normally distributed scores (see Figure 1.1).

FIGURE 1.1: THE BELL CURVE (I.E., THE CURVE OF NORMALLY DISTRIBUTED SCORES)

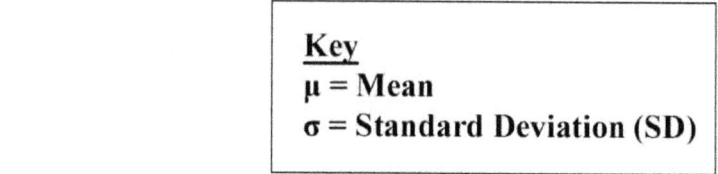

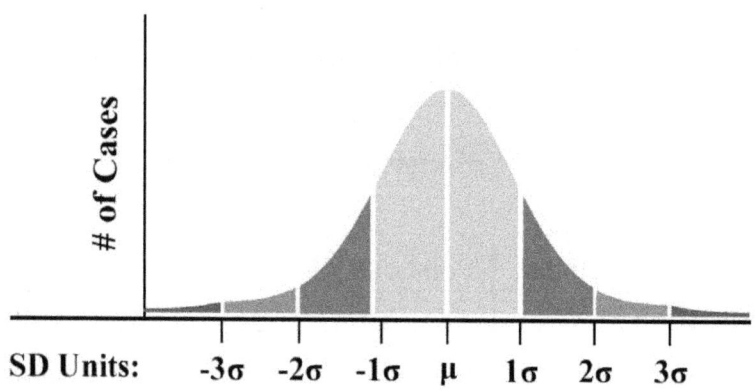

© Lisa A. Cottone

The bell curve is a graphical representation of all large, hypothetical distributions of data. In theory, the tails on the left and right sides of the curve extend outward to infinity, as the theory from which the bell curve was derived assumes that all distributions are infinitely large. Interestingly, most large distributions – from human height, to the IQ score of 12-year olds in California, to the lifespan of dung beetles – tend to conform to a bell-shaped curve, with the largest percentage of scores near the middle of the distribution around the mean (symbolized in this graph by the Greek letter, μ, which is called Mu and pronounced "myoo"), and with diminishing percentages of scores extending outward from the middle (i.e., both above and below the mean).

As shown in Figure 1.1 (on the previous page) and Figure 1.2 (on the following page), extending outward from the mean (μ), there are three negative values on the left (-1σ, -2σ and -3σ) and three positive values on the right (1σ, 2σ and 3σ). These values refer to the standard deviation units of the distribution, and the Greek letter σ (which is called sigma) is used to represent a standard deviation unit.

Specifically, 1σ (i.e., 1 sigma) refers to a value on the graph that is equal to the mean plus 1 standard deviation unit. In the example with Mr. Hardnose's class, where the mean was **68** and the standard deviation was **11.88**, the point on the graph corresponding to 1σ would be equal to a raw score of **79.88** (i.e., **68 + 11.88**). This is illustrated in Figure 1.2.

Similarly, using the data from Mr. Hardnose's class, 2σ would be equal to the mean plus 2 standard deviation units (i.e., **68 + 11.88 + 11.88**) which sums to **91.76**; and 3σ would be equal to the mean plus 3 standard deviation units (i.e., **68 + 11.88 + 11.88 + 11.88**) which sums to **103.64**.

On the left side of the graph, -1σ refers to a value that is equal to the mean minus 1 standard deviation unit. Hence, the point on the graph corresponding to -1σ would be equal to **56.12** (i.e., **68 - 11.88**). Going down the line, -2σ would be equal to the mean minus 2 standard deviation units (i.e., **68 – 11.88 – 11.88**) yielding a value of **44.24**; and -3σ would be equal to the mean minus 3 standard deviation units (i.e., **68 - 11.88 - 11.88 - 11.88**) yielding a value of **32.36**.

FIGURE 1.2: FINAL GRADES FOR MR. HARDNOSE'S CLASS, PLOTTED ON THE BELL CURVE

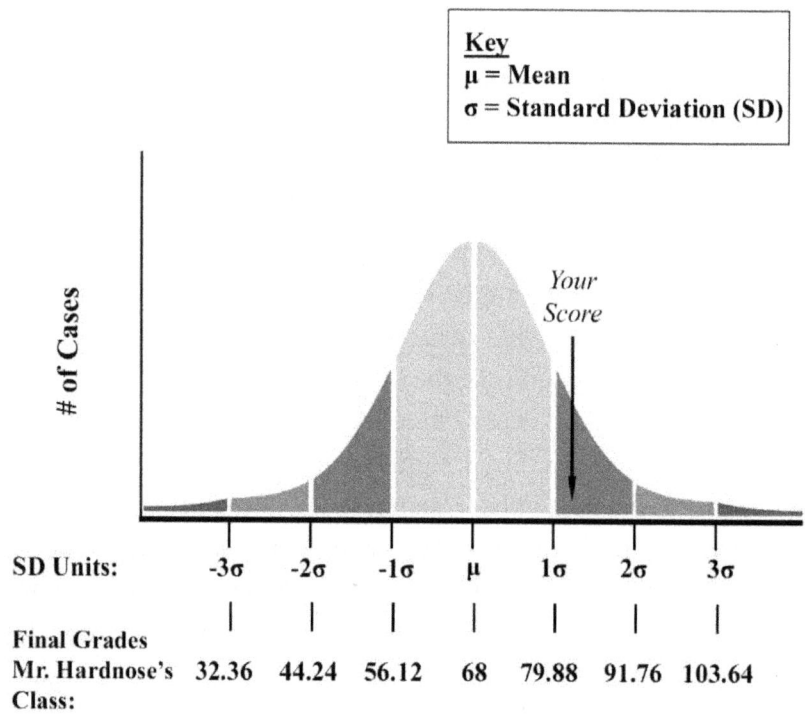

© Lisa A. Cottone

As noted above, the theory from which the bell curve was derived assumes that all distributions are infinitely large, and therefore, in theory, this graph could extend outward infinitely on both sides, to -∞ σ on the left, and +∞ σ on the right. However, since it is rare for scores in most real distributions of data to extend beyond -3σ and 3σ, these are often the highest values displayed on bell curve graphs.

Now, the reason why it is important for us to explain the values on the bell curve is so that you can understand exactly what a z-score refers to, which is the number of standard deviation units a given score is above or below the mean. Hence, a z-score of 1.0 is equal to 1σ (i.e., the mean plus one standard deviation unit); a z-score of 2.0 is equal to 2σ (i.e., the mean plus two standard deviation units); a z-score of -1.0 is equal to -1σ (i.e., the mean minus one standard deviation unit); and a z-score of -2.0 is equal to -2σ (i.e., the mean minus two standard deviation units); and so on, and so forth…

Most of the time, z-scores for a given distribution do not conform neatly to whole values, like 1.0, 2.0, or 3.0; however, if you understand the concept, interpreting actual z-scores is still quite easy and straightforward. For instance, if we consider the z-score for your final grade in chemistry (as shown in Table 1.2), which was **1.26**, this means that your grade was 1.26 standard deviation units above the mean of Mr. Hardnose's class. Interpreting z-scores is as easy as **π,** right?

Sibling Rivalry: Z-scores & Percentiles

As alluded to earlier, z-scores are just one type of standardized score that provides useful information about a particular data point relative to other data points in the distribution from which it came. Another type of standardized score – one with which many people are more familiar – is the percentile. In many instances percentiles can be used interchangeably with z-scores, and each can be converted to the other. Unlike z-scores, which, in theory, can range from -∞ to +∞, percentiles[6] have a restricted range of 0 to 100. In simplest terms, a given percentile represents the value within a distribution, below which a certain percentage of scores fall. Hence, if an individual achieves a percentile of 84, this means his score is higher than 84% of the other individuals in the same distribution.

Perhaps the context with which most people are familiar with percentiles is the SAT (Scholastic Assessment Test). Though the

---

[6] Percentiles are not limited to whole numbers, so, in theory, there are an infinite number of percentiles within this finite range of 0 to 100.

current version of the SAT exam, introduced in 2005, has three sections of performance (Mathematics, Critical Writing, and Critical Reading), with possible scores on the full exam ranging from 600 to 2400, the version of the exam that you may be most familiar with had two sections (Mathematics and Verbal), with possible scores on the full exam ranging from 400 to 1600. From year to year, changes in the difficulty of the exam, scoring procedures and various other anomalies meant that the mean of each cohort would vary, in some cases by more than 30 points. Hence, traditional scores on the SAT alone could not be trusted to give an accurate depiction of an individual's performance on the exam. As such, percentile scores are used because they provide information about how a student performed on the SAT exam relative to his or her peers taking that exact same exam, in a way that controls for factors like exam difficulty and other year-to-year differences. Hence, if you check your SAT transcript from high school, the percentile of your scores will tell you not only how you did on the exam relative to your peers, but how you might have done on the current exam with its new scoring system. If your SAT score yielded a percentile of 50, that means that your score was higher than 50% of your peers taking the same exam that year. Your performance would thus be equivalent to someone achieving a score at the 50th percentile on today's SAT exam, even though the raw scores on the respective exams may differ. For example, a score at the 50th percentile in 1996 for the full exam would have been 1013, whereas on the 2011 exam, a score at the 50th percentile would be 1500.[7] In Chapter 8, we provide an expanded discussion of the SAT exam and how z-scores could be used to make more sophisticated, and possibly fairer, comparisons between students taking the exam.

    For the most part, percentiles and z-scores provide the same type of information, though on different continuums. Whereas the

---

[7] National Center for Education Statistics (NCES; nces.ed.gov); SAT mean scores of college-bound seniors and percentage of graduates taking SAT, by state; extracted June 20, 2013.

mean of a distribution yields a z-score of 0, and the standard deviation points emanating from the mean extend outward in positive and negative directions at fixed units in multiples of 1.0 (i.e., 1.0, 2.0, 3.0, etc.), when it comes to percentiles, that same exact mean translates into a percentile of 50, and the standard deviation points emanating from that mean extend outward in positive and negative directions at variable units that decrease the further they are from the mean. On the following page, in Table 1.4, we list the percentiles at various standard deviation units emanating from the mean, along with their z-score equivalents. This is also illustrated graphically in Figure 1.3. As you can see, the mean of a distribution always corresponds to a z-score of 0, and a percentile of 50.

TABLE 1.4: Z-SCORES & PERCENTILES DIRECTLY COMPARED

STANDARD DEVIATION UNITS FROM THE MEAN	Z-SCORE	PERCENTILES
-5.0 σ	-5.0	< 0.001
-4.0 σ	-4.0	0.003
-3.0 σ	-3.0	0.135
-2.0 σ	-2.0	2.275
-1.0 σ	-1.0	15.865
0 σ (a.k.a. the mean)	0 (a.k.a. the mean)	50 (a.k.a. the mean)
1.0 σ	1.0	84.135
2.0 σ	2.0	97.725
3.0 σ	3.0	99.865
4.0 σ	4.0	99.997
5.0 σ	5.0	> 99.999

FIGURE 1.3: Z-SCORES & PERCENTILES PLOTTED ON THE BELL CURVE UP TO ±5σ

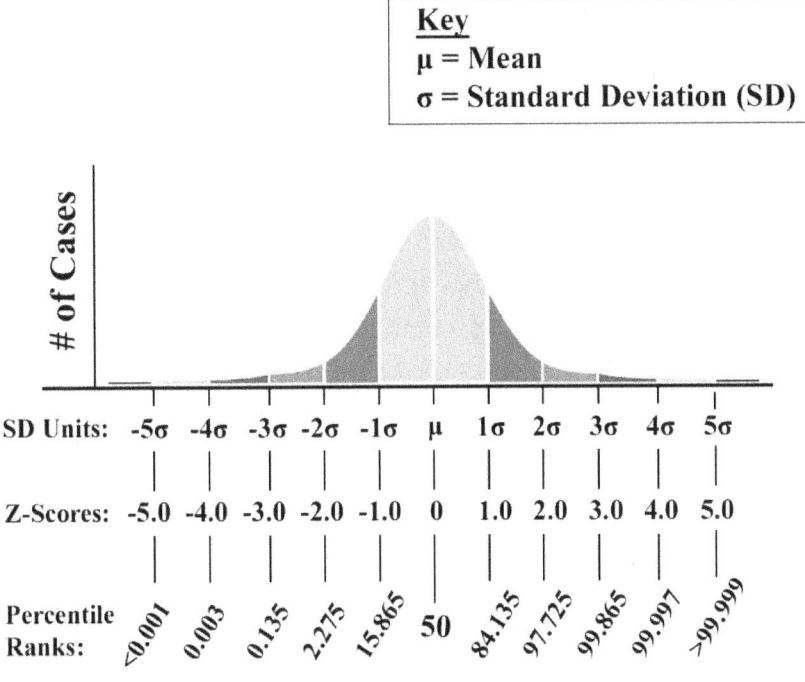

© Lisa A. Cottone

At this point, an obvious question might be: "If percentiles are statistically equivalent to z-scores, and people are more familiar with them, why not just use percentiles for the statistical comparisons of this book instead of z-scores?" The reason we chose to use z-scores instead of percentiles is because we believe that for many readers who may be unfamiliar with standardized scores, it is easier to recognize the statistical rarity of exceptional achievements on the z-score continuum (ranging, in theory, from -∞ to +∞) than on the percentile continuum (ranging from 0 to 100).

For instance, let's say we wanted to compare the standardized scores for height for a man who is 6'6" tall with a man who is 7'0" tall (see Table 1.5 for an expanded presentation of height and associated z-score values). A man who is 6'6" has a height that is 3σ (i.e., 3 standard deviation units above the mean for height, which in this case, is 5'10"), and a man who is 7'0" has a height that is 5σ (i.e., 5 standard deviation units above the mean for height). Converting these values to z-scores and percentiles, the man who is 6'6" has a z-score for height of 3.0 and the percentile for that height is 99.87 (i.e., this man is taller than 99.87% of other men); the man who is 7'0" has a z-score for height that is 5.0 and the percentile for that height is 99.99 (i.e., he is taller than 99.99% of other men). Now, if we are to compare the percentiles of the two men's heights – 99.87 to 99.99 – it doesn't appear that there is a tremendous difference (0.12 percentile units), because percentiles asymptote to 100 (i.e., they get closer and closer to 100 without ever reaching 100); however, if we compare the *z-scores* for the two men – 3.0 to 5.0 – this difference appears much more substantial (2.0 z-score units) because, in theory, z-scores extend outward to infinity. Hence, in order to make more salient the magnitude of the difference between two outstanding scores, we believe that z-scores are much better at doing this than percentiles. So, while the big brother of the standardized score family, the percentile, usually gets all the attention and fame, in our book, little brother z-score has exactly what we are looking for. Perhaps these two siblings have some family therapy in store for their future.

Z-scores in the Real World

The contexts with which z-scores are primarily used today are in the fields of clinical neuropsychological testing and research. Neuropsychological testing is often provided to kids when there is a suspicion that a learning disability is present. It is also provided to adults to check for the presence of any neurological or neuropsychological deficits, particularly following traumatic brain injury, onset of a neurological disease, or if dementia is suspected. A typical neuropsychological battery consists of dozens of tests, across myriad domains of functioning including: verbal and nonverbal intelligence; academic achievement; verbal and nonverbal memory; speech production; attention; gross motor skills; fine motor skills; reading fluency; reading comprehension; communication fluency; inhibition; abstract reasoning, and many others. Each of the tests given in a typical neuropsychological examination is scored differently. Some have a range of 40 to 160 points, with the mean being 100 points; others might have a range of zero to 60 points, with means varying by age and reference group; and still other tests may yield scores, not in the form of points, but in the form of time-to-completion, ranging from zero seconds to infinity. For example, on a typical neuropsychological profile, you may need to compare a patient's Full Scale IQ score of 105 (out of 160) on the SB5 (Stanford-Binet Intelligence Scales, Fifth Edition) to his score of 50 (out of 60) on the Boston Naming Test (a test of visual recognition and verbal fluency) and his score of 82 seconds (for completion) on the Grooved Pegboard Test (a test of fine motor skills).

In short, a neuropsychological exam requires a clinician to compare apples to oranges, not just in terms of the actual skills being assessed, but also in terms of the types of scores being compared. It is as if you need to compare a baseball player's batting average to the number of goals a hockey player scored, to the time it takes a sprinter to finish the 100-yard dash. How can this be done? Standardized scores, of which percentiles and z-scores are just a few, assist in this process. For reasons that are more complex than what is necessary to explain here, in neuropsychological testing, percentiles are more commonly used than z-scores to assist in these comparisons. However, in research (including research on neuropsychological test data), z-scores tend to be more commonly used. Having experience and training in both clinical and research settings, we believe that for the purposes of this book, z-scores will illuminate key distinctions between superb performances with greater clarity than will percentiles.

In an effort to help you understand how z-scores relate to familiar concepts in the real world, Table 1.5 lists a range of z-scores (from -5 to +5) for two variables with which most people have some experience: human height and IQ. The data for height in this table come from the American National Health and Nutrition Examination Survey, which reports American men as having a mean height of 70" (i.e., 5'10" tall) with a standard deviation of 2.8", and American women as having a mean height of 65" (i.e., 5' 5" tall) with a standard deviation of 2.5". For IQ, on most standardized, normed IQ tests, the mean is usually set at 100 and the standard deviation is set at 15. As a frame of reference for IQ, acceptance into Mensa International, the society for individuals with high IQs, requires an IQ score in at least the 98th percentile (or about 2σ)[8], which equates to an IQ score of about 130 on the most common intelligence tests used today.

As you can see from Table 1.5, an IQ score of 130, which will get you in to Mensa, yields a z-score of 2.0, which is pretty impressive. But just how impressive is a z-score of 2.0? What would that equate to in human height? Well, by looking at Table 1.5 you can determine that a z-score of 2.0 would be statistically equivalent to a man who is 6 feet, 3.6 inches tall, and a woman who is 5 feet, 10 inches tall. If you can understand this, pat yourself on the back: you have just passed our first exam and now you know how to compare *apples to oranges*!

---

[8] http://www.mensa.org/about-us

TABLE 1.5: Z-SCORES OF ADULT HUMAN HEIGHT & IQ

Z-SCORE	IQ[9]	ADULT MALE HEIGHT[10]	ADULT FEMALE HEIGHT[7]
-5.0	25	56.0" (4 ft. 8.0 in.)	52.5" (4 ft. 4.5 in.)
-4.0	40	58.8" (4 ft. 10.8 in.)	55.0" (4 ft. 7.0 in.)
-3.0	55	61.6" (5 ft. 1.6 in.)	57.5" (4 ft. 9.5 in.)
-2.0	70	64.4" (5 ft. 4.4 in.)	60.0" (5 ft. 0 in.)
-1.0	85	67.2" (5 ft. 7.2 in.)	62.5" (5 ft. 2.5 in.)
0	**100**	**70" (5 ft. 10.0 in.)**	**65.0" (5 ft. 5.0 in.)**
1.0	115	72.8" (6 ft. 0.8 in.)	67.5" (5 ft. 7.5 in.)
2.0	130	75.6" (6 ft. 3.6 in.)	70.0" (5 ft. 10 in.)
3.0	145	78.4" (6 ft. 6.4 in.)	72.5" (6 ft. 0.5 in.)
4.0	160	81.2" (6 ft. 9.2 in.)	75.0" (6 ft. 3.0 in.)
5.0	175	84.0" (7 ft. 0 in.)	77.5" (6 ft. 5.5 in.)

---

[9] Esther Strauss, Elisabeth M. S. Sherman, Otfried Spreen (2006). *A Compendium of Neuropsychological Tests: Administration, Norms, and Commentary*.
[10] NHANES II Study (National Health and Nutrition Examination Survey of 1976-1980), as cited in *Seeing Through Statistics, 2nd Edition* by Jessica M. Utts (1999). Also found on the National Center for Health Statistics website: http://www.cdc.gov/nchs/nhanes.htm

From Apples and Oranges to Baseballs, Basketballs Pigskins and Pucks: How *Our* Z-scores Were Obtained

All of the sports data used in our analyses were acquired from sources we considered reputable. Our baseball data were obtained from Major League Baseball – specifically, from MLB.com, while the data for football, basketball and hockey, reported in Chapter 6, were obtained from Sports-Reference.com. We obtained our data from these sources, not only because we trusted them to have the most accurate data, but also because they each have a powerful sorting feature, allowing one to select player data that meet specific criteria of interest. For our analyses, we sorted and selected cases according to various criteria so that we could make the most refined and meaningful comparisons between players.

Before extracting any statistics, we debated how we would approach the analysis of such a large pool of data in a way that was both interesting and efficient. Rather than computing z-scores for every player, in every season, for every available statistical category (a monumental task, even for those with supercomputers and a lust for solitude), we decided to take a more streamlined approach. In our streamlined approach, we chose to compute z-scores only for the players who led certain statistical categories that we deemed most interesting, and only for selected seasons.

For baseball, we chose to examine a mix of traditional and sabermetric statistics in both hitting and pitching domains. Within the hitting domain, the traditional statistics we examined were batting average, home runs and RBI[11] and the sole sabermetric statistic we examined was OPS.[12] For pitching, the traditional statistics we examined were wins, ERA[13] and strikeouts, while the

---

[11] RBI = Runs Batted In.
[12] OPS = On-base [percentage] Plus Slugging [average]
[13] ERA = Earned Run Average

sole sabermetric statistic we examined was WHIP.[14] We then identified seven major eras of baseball history, demarcated by the popular reference book *The Baseball Timeline*[15] (an official publication of Major League Baseball). According to *The Baseball Timeline*, the seven eras of baseball history are:

1. 1876 - 1900 – Pre-Modern Era[16]
2. 1901 - 1919 – Deadball Era[17]
3. 1920 - 1960 – Home Run Era
4. 1961 - 1975 – Expansion Era
5. 1976 - 1985 – Free Agency Era
6. 1986 - 2007 – Steroid Era
7. 2008 – 2014[18] – Post-Steroid Era

For each of the seven eras listed above, we only calculated z-scores for the players who had the best single-season performance for the ***entire era***, within their ***respective league*** (i.e., American or National), for the ***selected statistics of interest*** mentioned above. For instance, in the Deadball Era, which spanned the years 1901-1919, we didn't calculate a z-score for home runs for *every player* across *every season* between 1901 and 1919: we only calculated z-scores for the home run totals of Babe Ruth in 1919 and Gavvy Cravath in 1915 because their single-season home run totals – 29 for Ruth, and 24 for Cravath – were higher than those of any other player in their respective leagues, for any season within the Deadball

---

[14] WHIP = Walks [plus] Hits [divided by] Innings Pitched
[15] Solomon, Burt (2001). *The Baseball Timeline.* DK Publishing, Inc. New York
[16-17] Though *The Baseball Timeline* set the years of the Pre-Modern Era as being between 1876-1902, and the years of the Deadball Era as being between 1903-1919, we decided to make 1900 the end of the Pre-Modern Era and 1901 the start of the Deadball Era for the purposes of our analyses, given that the American League was founded in 1901.
[18] 2014 is the year in which this book was completed. Hence the 2014 baseball season is the final season in which statistics were obtained for the data analyses of this book.

Era. Likewise, in all other eras, and for all other statistical categories, we only computed z-scores for the single-season category leaders (i.e., for batting average, home runs, RBI, OPS, wins, ERA, strikeouts, and WHIP) of that particular *era*, within each league.

As briefly noted above, we examined the data for each league separately. The reason is because for much of baseball history, the leagues were very different, and still are to some extent. Each league has its own distinct ballparks and even to this day most of the teams in one league still do not play any games in the ballparks of the teams from the other league. In addition, the leagues have different rules regarding the designated hitter (since 1973), and for many years there were even different umpires and umpiring norms for each league (until umpires began working both leagues during the 2000 season). Furthermore, each league had a different timeline for integrating non-White players. Hence, in light of all of these differences, we decided to look at the data for the American League and the National League separately.

Additionally, it should be noted that when computing z-scores for each category leader, rather than using means and standard deviations derived from the entire pool of players within a respective league for a given season, we narrowed the pool (using the sorting feature on MLB.com) for each season of interest to include only the players who qualified for the batting title[19] (for

---

[19] Leaderboard Glossary – Baseball. www.Baseball-reference.com. Retrieved 2012-05-26:

Qualifications for the Batting Title

- Pre-1920: A player generally had to appear in 100 or more games when the schedule was 154 games; and 90 games when the schedule was 140 games.
- 1920–1949: A player generally had to appear in 100 games to qualify in the National League; the American League used 100 games from 1920–1935, and 400 at-bats from 1936–1949.
- 1950–1956: A player needed 2.6 at-bats per team game originally scheduled. (With the 154-game schedule of the time, that meant a rounded-off 400 at-bats.) From 1951–1954, if the player with the highest

hitting statistics) and the ERA title[20] (for pitching statistics) within that season for each respective league. We did this to avoid skewing

> average in a league failed to meet the minimum at-bat requirement, the remaining at-bats until qualification (e.g., five, if the player finished the season with 395 at-bats) were hypothetically considered hitless at-bats; if his recalculated batting average still topped the league, he was awarded the title. This standard applied in the AL from 1936–1956.
> - 1957 – The Present: A player needs 3.1 plate appearances per team game originally scheduled; In the 154-game schedule, the required number of plate appearances was 477, and since the advent of the 162-game schedule, the requisite number of plate appearances has been 502. (Adjustments to this 502 figure have been made during strike-shortened seasons, such as 1972, 1981, 1994, and 1995.)
> - *From 1967 to the present, if the player with the highest average in a league fails to meet the minimum plate-appearance requirement, the remaining at-bats until qualification (e.g., five at-bats, if the player finished the season with 497 plate appearances) are hypothetically considered hitless at-bats; if his recalculated batting average still tops the league, he is awarded the title.*

[20] Article by Dan Levitt in Society for American Baseball Research's *The National Pastime: No. 25.*

Qualifications for the ERA Title
- In 1917, the National League adopted the standard that a pitcher had to have pitched a minimum of 10 complete games in order to qualify as an ERA leader. Before that, the National League used a variety of standards: in 1912, it was 15 games pitched; in 1913, it was five complete games; from 1914-1915, it was 15 games pitched; and in 1916, it was 12 games pitched.
- In 1946, the American League adopted the standard that a pitcher had to have pitched a minimum of 10 complete games. Before that, the league used various criteria. Up until 1919, the American League simply listed the top pitchers by ERA without identifying a standard. In 1919, the American League finally adopted a minimum standard of 45 innings pitched. The next year, 1920, it changed the standard to 10 complete games, but in 1921, it reverted to 45 innings pitched. The following year, 1922, it switched back to 10 complete games, where it remained for three seasons. In 1925, the American League once again reverted back to a standard of 45 innings pitched.
- In 1951, the standard of the Major League was changed to one inning pitched per team game, which is the current standard (i.e., 162 innings

the means and standard deviations for each season, which would have happened if we included, for example, a player who batted .500 in a given season, but only had 10 at-bats; or, alternately, a pitcher who may have had an ERA of 0.00 but pitched only one inning. Here it should be noted that relief pitchers were not included in the main set of our analyses for pitching (as they generally do not qualify for the ERA title, according to the norms of Major League Baseball). However, later in the book in Chapter 4 we provide z-scores for relief pitchers who won the MVP and in that chapter we describe in detail the methods we used to classify pitchers (as either a starter or reliever) and compute their z-scores.

For each of the other sports we examined – football, basketball and hockey – we made similar decisions to pare down the qualifying pool so as to avoid the inclusion of part-time players into our analyses. Since baseball tends to have more formal rules for inclusion into the pool of full-time players for various statistical distinctions, such as the batting title and ERA title, we used baseball as a guide to help us determine the appropriate inclusion criteria for the other sports. As such, the inclusion guidelines that we chose for the other sports generally corresponded to the least restrictive qualification standards for baseball's batting title in the history of the sport: namely, participation in at least 65% of the team's games.[21] Regarding the specific statistical categories of interest for the other three sports, we examined the following:

---

pitched for a standard 162-game season, without any rainouts or tie-breaker playoff games).

[21] Used between 1920 and 1949 in the National League, when the guideline for qualification for the batting title was playing in 100 of the team's 154 games (i.e., 65%).

- **Football**
    - Offense: *Rushing Yards; Touchdown Passes; Completion Percentage; Receiving Yards;* and *Touchdown Receptions.*
    - Defense: *Sacks;* and *Interceptions.*

- **Basketball**
    - Scoring: *Total Points; Points Per Game (PPG); Total Rebounds; Rebounds Per Game (RPG); Total Assists; Assists Per Game (APG);* and *Three-Point Field Goal Percentage.*

- **Hockey**
    - Scoring: *Total Goals; Total Assists;* and *Total Points.*
    - Goaltending: *Wins; Shutouts; Goals Against Average;* and *Save Percentage.*

As noted above, we didn't compute z-scores for the selected statistics for *every player*, in *every season*. Rather, as a means of making certain specific comparisons across sports, we computed z-scores for the single-season record holders in each category of interest, within each sport. We then pitted these z-scores in direct competition with those we derived from our baseball analyses for a **BATTLE ROYAL** of z-scores to identify the best achievements of the four major American team sports. Who emerges from this battle royal as the Heavyweight Champion of Z-scores? Read on to find out…

The Least You Need to Know

1. Z-scores are a standardized statistic, like percentiles, that indicate how a person's performance on a given measure compares to others in the same group.

2. Virtually any measure of performance, including most sports statistics, can be converted to z-scores.

3. The formula to compute a z-score is:

$$\text{Z-score} = \frac{(Score\ X - Mean)}{Standard\ Deviation}$$

4. For the purposes of this book, the better the performance, the higher the z-score in the positive direction (e.g., for home runs a z-score of +2 is better than a z-score of +1, 0, -1, -2, etc.).

NOTE: Technically, for variables in which lower values represent better performance (e.g., ERA and WHIP) the more _negative_ the z-score, the better the performance. However, to simplify presentation in this book, negative z-scores for ERA and WHIP were flipped to positive z-scores to help readers make easier comparisons with z-scores from other statistical categories.

The Least You Need to Know

5. Though, in theory, z-scores extend outward to infinity in both positive and negative directions (i.e., they range from $-\infty$ to $+\infty$), for most variables in the real world it is rare to encounter z-scores extending beyond -5 and +5, and the vast majority of z-scores fall between -2 and +2. The average or "mean" score for any statistic will always have a z-score of 0.

6. As a familiar point of reference, an IQ score of 130, which is the minimum IQ necessary to get into Mensa (the society for people with high IQs), is equivalent to a z-score of 2.0. A z-score of +5.0 in the realm of height among men yields a measurement of 7'0" (which is about the height of Shaquille O'Neal).

Chapter 2

Runs, Hits & Eras:
A Summary of Baseball History by Era

"America has rolled by like an army of steamrollers. It has been erased like a blackboard, rebuilt and erased again. But baseball has marked the time."

- Terence Mann,
Field of Dreams

At nearly 150 years old, baseball is at once a revered American institution and also a never-ending story that is still being written. Like the Bible, baseball's history can be divided into many books and chapters and organized in countless ways according to each observer's personal interest and background. For the purposes of this book, and for the benefit of our statistical analyses, we have conceived of baseball history as being divided into seven distinct eras. Spanning the period from the mid-19th century to the present, each so-called "era" features its own quirks, trends and rule changes. When woven together, they create a map of our national pastime and give life to the z-scores[22] presented in subsequent chapters.

In this chapter we summarize each of the seven eras of baseball history listed in Chapter 1, noting the factors that made each respective era unique, as it relates to changes in rules, ballparks, equipment, cultural norms, player demographics and other nuances. We also identify the category leaders within each era, for each league, for each of the selected statistics of interest: batting average, home runs, RBI and OBP, among hitting categories; and wins, ERA, strikeouts and WHIP, among pitching categories. Z-scores for the category leaders identified in this chapter will then be reported in Chapter 3.

---

[22] For most of the z-scores we computed, better performance was associated with a higher, more positive z-score, because for most statistics (e.g., home runs) higher raw scores indicate better performance. However, for two of the pitching categories – ERA and WHIP – better performance is associated with lower raw scores (e.g., an ERA of 1.50 is better than an ERA of 4.50). As such, when we computed z-scores for these categories, the best performances yielded negative z-scores (e.g., Bob Gibson's ERA in 1968 yielded a z-score of -2.76). However, for ease of comparison with the z-scores of other statistical categories, we made a simple correction: we inverted the negative numbers into positive numbers (e.g., -2.76 became 2.76). Hence, for those categories, the actual z-scores will be positive, rather than negative.

Pre-Modern Era (1876-1900)

To keep with official records, our story starts in 1876, with the birth of the National League, though the game's origins actually stretch as far back as 1845. It was in this year that Alexander Cartwright presented the first official rules of the game, which included references to strikeouts, equidistant bases and player interference. In 1857, the first formal baseball convention was held in New York City, and it was here that team representatives agreed that all games would last nine innings and could end after five innings in cases of inclement weather. In 1858, it was ruled that a ball must be caught on the fly in order to be counted as an out, and four years later, runners were required to touch all bases in order to score.

Professional baseball began in 1869, when the Cincinnati Red Stockings became the nation's first "pay-for-play" organization. But it wasn't until the first structured league emerged seven years later that universal rules were put into effect. Pitchers were not allowed to release the ball above their waists and the mound was designated at 45 feet from home plate (though the distance was pushed back to 50 feet by the 1882 season). In 1884, overhand pitching was legalized and the schedule expanded to 112 games from the original slate of 70.

A pair of unique caveats debuted for the 1887 season, only to be abolished the following year: four strikes were required for an out and all walks counted as hits. In 1888, three strikes for an out and modern walk rules became the norm. One of the final Pre-Modern Era rule changes took effect in 1895, when maximum bat diameters were increased from two-and-a-half inches to two-and-three-quarter inches.

Schedules were once again expanded in 1893, this time growing to between 140 and 154 games. Moreover, in 1884, the mound's distance from home plate was increased to the current 60 feet 6 inches.

Pitching rotations were vastly different than what they are now. Rather than the five-man rotations that we have today, one-to-two-man rotations were popular from the early 1870s to late 1880s, while three-man rotations were more the fashion in the 1890s. With the advent of the Modern Era, in 1901, four-man rotations started to become more popular, and this was the case for much of the 20th century until the early 1990s, when five-man rotations became the norm.

An understanding of the change in size of pitching rotations is important for many reasons, and particularly for statistical comparisons. Each time pitching rotations increased in size, starting pitchers had fewer and fewer starts per season and thus fewer opportunities to win games and record strikeouts. For much of baseball's history, the arbitrary benchmark of winning 20 games has been the standard by which starting pitchers are judged. Winning 20 games, a pitcher is deemed to have had a great season. But winning 20 games has become harder over the years as pitching rotations expanded and individual pitchers have fewer starts. With today's five-man pitching rotation, it is 20% harder to win 20 games than it was for much of the 20th century because starting pitchers have 20% fewer starts. The same can be said for all other cumulative pitching statistical benchmarks, such as the benchmark of recording 300 strikeouts in a season.

In addition to the rule changes and evolution of norms discussed above, it is also important to note that for most of the players of the Pre-Modern Era, playing baseball was only a part-time job, as players typically held other full-time jobs in the off-season (usually involving hard physical labor), and other part-time or full-time jobs during the baseball season. As a result, players' conditioning, energy and physical performance levels were likely compromised, especially in comparison to today's players, whose lives revolve around preparing – physically *and* mentally – for their primary job of playing baseball for 162 games per year. Professional baseball players had day jobs and off-season jobs well into the 20th century – even as late as the 1940s and 1950s.

From a scorekeeping perspective, one major anomaly of this era – walks counting as hits – made for a favorable environment for inflated batting averages, and this was the case for the 1887 season.

TABLE 2.1: CATEGORY LEADERS IN THE PRE-MODERN ERA (1876-1900)[23]

HITTING STATISTICS

NATIONAL LEAGUE

STATISTIC	PLAYER	YEAR	TEAM	RAW VALUE	LEAGUE MEAN	LEAGUE SD[24]
BATTING AVERAGE	Hugh Duffy	1894	Boston Braves	.440	.320	.04
HOME RUNS	Ned Williamson	1884	Chicago Cubs	27	4.57	5.69
OPS	Hugh Duffy	1894	Boston Braves	1.196	.850	.12
RBI	Sam Thompson	1887	Detroit Wolverines	166	64.94	26.40

PITCHING STATISTICS

NATIONAL LEAGUE

STATISTIC	PLAYER	YEAR	TEAM	RAW VALUE	LEAGUE MEAN	LEAGUE SD
ERA	Tim Keefe	1880	Troy Trojans	0.86	2.42	.83
STRIKEOUTS	Charley Radbourn	1884	Providence Grays	441	169.38	121.86
WHIP	Tim Keefe	1880	Troy Trojans	0.80	1.12	.21
WINS	Charley Radbourn	1884	Providence Grays	59	17.54	16.15

---

[23] Statistics for this era are only available for the National League. The American League was not a part of Major League Baseball at this time.
[24] SD – Standard Deviation

Deadball Era (1901-1919)

Our next era began in 1901, with the birth of the American League. The year is arbitrary, but the existence of two formal leagues provides a greater sample size of official records. As its name indicates, the Deadball Era was characterized by a severe lack of home runs, which was partially due to the use of a ball that was considered "dead" by today's standards. Though the early rules of baseball called for all balls to weigh between five and five-and-a-half ounces, with a circumference of between nine and nine-and-a-half inches, baseballs during much of this era (and the previous era) were typically larger and heavier. This prompted officials to introduce a standard ball with a cork center by 1911. This livelier baseball became known as the "rabbit ball," even though home run numbers remained stagnant in the years immediately following its introduction.

In addition to the use of a larger and heavier ball for much of this era, the lack of power hitting may also be attributed to several other factors. Perhaps the greatest hindrance to home runs during this period was the sheer magnitude of the ballparks. Most playing fields during this and the previous era featured center field fence distances *averaging* 490 feet; and in many cities, there were no home run fences, giving outfields nearly endless dimensions. The public record is incomplete, but available documents confirm colossal center field distances in at least nine cities. Chicago's West Side Grounds registered at 560 feet, while Hilltop Park in New York came in at 542 feet. However, no park came close to the Huntington Avenue Grounds in Boston, where 635 was the number to beat. These massive fields made it very challenging for hitters to achieve home runs.

As if the "dead" ball and cavernous ballparks weren't enough obstacles to home runs, in 1903, the Rules Committee decreed that the pitcher's mound could be as high as 15 inches, which gave pitchers an even more *elevated* advantage.

Though home runs were rare during the Deadball Era, they were not nonexistent. From 1901 to 1919, the home run leaders in both leagues struck an average of 12 long balls per season. Given the enormous dimensions of the ballparks, a large number of home runs during this era were of the *inside-the-park* variety, which, obviously, are more difficult to produce than a single swing sending the ball over a fence (in either left, center, or right field) between 300 and 425 feet away, which is the norm for home runs today. But even inside-the-park home runs were targeted for suppression by the end of the era, as it seemed that ground balls rolling deep into the outfield posed too much of a nuisance for outfielders. As such, an interesting rule change gives life to this argument: in 1914, the Rules Committee *restricted* a batter to three bases if a defensive player threw his cap or glove to stop a hit ball. Despite rules, balls and playing fields that seemed designed to suppress home runs, batting averages in the Deadball Era remained high (though not as high as in the Pre-Modern Era), as did RBI totals (batting average leaders in both leagues hit about .360 during this period).

As we shift our focus from hitters to pitchers, it is important to note that the norms for pitching rotations changed in the Deadball Era, with four-man rotations now becoming the norm (as opposed to the one-man, two-man and three-man rotations of the Pre-Modern Era). As such, starting pitchers had more opportunities to rest their arms, but they were also given fewer opportunities to win games and acquire strikeouts. This is why, among other reasons, the era's premiere pitchers deserve ample credit. And until 1921, when umpires began rubbing balls with mud to improve grip, balls were more difficult to throw with accuracy.

However, if ever there was a watershed season in baseball history, a year that would forever alter the game's landscape, our statistics point to 1919. During this season – one that can aptly be called the *Year of the Babe* – George Herman "Babe" Ruth slugged 29 home runs and posted a hefty 1.114 OPS. To put this achievement into its appropriate context, it is important to note that the average home run total that year in the American League was a mere 3.47, and the league average for OPS was a paltry .735. Single-handedly, the Sultan of Swat brought the game out of its doldrums, opening the floodgates to the Home Run Era.

TABLE 2.2A: CATEGORY LEADERS IN THE DEADBALL ERA (1901-1919)

HITTING STATISTICS

NATIONAL LEAGUE

STATISTIC	PLAYER	YEAR	TEAM	RAW VALUE	LEAGUE MEAN	LEAGUE SD
BATTING AVERAGE	Cy Seymour	1905	Cincinnati Reds	.377	.277	.03
HOME RUNS	Gavvy Cravath	1915	Philadelphia Phillies	24	3.82	4.21
OPS	Heinie Zimmerman	1912	Chicago Cubs	.989	.761	.08
RBI	Gavvy Cravath	1913	Philadelphia Phillies	128	57.60	20.38

AMERICAN LEAGUE

STATISTIC	PLAYER	YEAR	TEAM	RAW VALUE	LEAGUE MEAN	LEAGUE SD
BATTING AVERAGE	Nap Lajoie	1901	Philadelphia Athletics	.426	.294	.03
HOME RUNS	Babe Ruth	1919	New York Yankees	29	3.47	4.48
OPS	Babe Ruth	1919	New York Yankees	1.114	.735	.106
RBI	Frank Baker	1912	Philadelphia Athletics	130	64.83	25.67

TABLE 2.2B: CATEGORY LEADERS IN THE DEADBALL ERA (1901-1919)

PITCHING STATISTICS

NATIONAL LEAGUE

STATISTIC	PLAYER	YEAR	TEAM	RAW VALUE	LEAGUE MEAN	LEAGUE SD
ERA	Tim Keefe	1880	Troy Trojans	0.86	2.42	.83
STRIKEOUTS	Charley Radbourn	1884	Providence Grays	441	169.38	121.86
WHIP	Tim Keefe	1880	Troy Trojans	0.80	1.12	.21
WINS	Charley Radbourn	1884	Providence Grays	59	17.54	16.15

AMERICAN LEAGUE

STATISTIC	PLAYER	YEAR	TEAM	RAW VALUE	LEAGUE MEAN	LEAGUE SD
ERA	Dutch Leonard	1914	Boston Red Sox	0.96	2.65	.56
STRIKEOUTS	Rube Waddell	1904	Philadelphia Athletics	349	129.51	58.39
WHIP	Walter Johnson	1913	Washington Senators	0.78	1.24	.14
WINS	Jack Chesbro	1904	New York Highlanders	41	16.03	7.30

Home Run Era (1920-1960)

What characterized the Home Run Era more than anything else was the number of Hall of Fame-caliber players who dominated the game during this period. This was the time when the titans of the sport built their legacies, and immortals like Babe Ruth, Joe DiMaggio, Ted Williams, Willie Mays and Stan Musial became national heroes. Romantics and traditionalists often refer to this era as baseball's Golden Age.

The Home Run Era loosely began in 1920, which was Babe Ruth's first season with the New York Yankees: a year after he broke the previous major league home run record (set by Ned Williamson in 1884) as a member of the Boston Red Sox. In between the 1919 and 1920 seasons, Ruth was famously sold to the Yankees from the Red Sox, and in 1920, he wasted no time in chopping away at *his own* home run record, smacking 54 home runs along with 137 RBI. He boasted an .847 slugging percentage. His gargantuan clout even prompted speculation of a livelier baseball, with some believing that the ball was purposely modified to take advantage of Ruth's rising popularity. Others believe that the ball itself did not change during this era, but that the game's power explosion was caused by two new rules implemented in 1920: old balls had to be removed from play when they became tattered, worn or dirty (which would make them both soft and difficult for hitters to see); and spitballs were no longer allowed, despite the fact that they were popular and effective at stopping batters from getting hits.[25]

Home runs exploded during this era even though new regulations mandated that the minimum distance for home run fences be pushed back in ballparks to at least 250 feet. Before this change in 1925, the minimum distance for home runs was 235 feet. Later during this era, new regulations mandated that all fields built after June 1, 1958, were required to provide minimum distances of

---

[25] *Evolution of the Ball.* Books. www.Google.com. July 1963. Retrieved October 7, 2011.

325 feet along the right and left field foul lines, and 400 feet to center, though existing ballparks with smaller dimensions were grandfathered in and allowed to stand as they were – something that players like the light-hitting Red Sox shortstop Johnny Pesky[26] were undoubtedly very happy about. But while the minimum distances for home runs increased, the oversized ballparks of the previous era (some, like the Polo Grounds and Forbes Field, with fences in center field and the power alleys exceeding 450 feet) were no longer in style, and the majority of ballparks were smaller, which made it easier for batters to hit home runs.

Ruth was the Home Run Era's superstar. He led the American League in homers, RBI and slugging percentages six times between 1919 and 1928. Other prominent American League sluggers included Lou Gehrig, Jimmie Foxx and Hank Greenberg, while Mel Ott, "Hack" Wilson and Chuck Klein powered their way through the National League.

As the game's offense grew, so did its popularity. Average attendance figures hovered near 10 million for most of the 1920s, and eventually passed that mark in 1930. But during the Great Depression and World War II, attendance dropped by 40 percent and didn't return to pre-Depression levels until after the war. In addition to losing fans and revenue, World War II also robbed baseball of the prime years of some of its most iconic players, including Ted Williams, Joe DiMaggio, Bob Feller, Stan Musial and Enos Slaughter, among others. When the war ended, the game exploded and league-wide attendance jumped to more than 20 million by 1948. Jackie Robinson famously opened the door for Black and

---

[26] The right field foul pole at Fenway Park is called "the Pesky Pole," named after Johnny Pesky, who hit several important home runs just inside the pole during his career (from 1942 to 1954). Pesky was not known to be a power hitter (he only had 17 career home runs), but the short distance between the pole and home plate (302 feet) allowed him to a slug a few round-trippers at Fenway Park, which was built in 1912.

Hispanic players. The Yankees and Dodgers grew into Major League Baseball's premier franchises.

TABLE 2.3A: CATEGORY LEADERS IN THE HOME RUN ERA (1920-1960)

HITTING STATISTICS

NATIONAL LEAGUE

STATISTIC	PLAYER	YEAR	TEAM	RAW VALUE	LEAGUE MEAN	LEAGUE SD
BATTING AVERAGE	Rogers Hornsby	1924	St. Louis Cardinals	.424	.303	.04
HOME RUNS	Hack Wilson	1930	Chicago Cubs	56	14.14	12.42
OPS	Rogers Hornsby	1925	St. Louis Cardinals	1.245	.837	.11
RBI	Hack Wilson	1930	Chicago Cubs	191	90.84	34.20

AMERICAN LEAGUE

STATISTIC	PLAYER	YEAR	TEAM	RAW VALUE	LEAGUE MEAN	LEAGUE SD
BATTING AVERAGE	George Sisler	1922	St. Louis Browns	.420	.303	.04
HOME RUNS	Babe Ruth	1927	New York Yankees	60	6.91	11.28
OPS	Babe Ruth	1920	New York Yankees	1.379	.795	.14
RBI	Lou Gehrig	1931	New York Yankees	184	82.87	32.65

TABLE 2.3B: CATEGORY LEADERS IN THE HOME RUN ERA (1920-1960)

PITCHING STATISTICS

NATIONAL LEAGUE

STATISTIC	PLAYER	YEAR	TEAM	RAW VALUE	LEAGUE MEAN	LEAGUE SD
ERA	Carl Hubbell	1933	New York Giants	1.66	3.16	.59
STRIKEOUTS	Dazzy Vance	1924	Brooklyn Dodgers	262	65.62	38.27
WHIP	Warren Hacker	1952	Chicago Cubs	0.95	1.30	.15
WINS	Dizzy Dean	1934	St. Louis Cardinals	30	15.24	5.00

AMERICAN LEAGUE

STATISTIC	PLAYER	YEAR	TEAM	RAW VALUE	LEAGUE MEAN	LEAGUE SD
ERA	Spud Chandler	1943	New York Yankees	1.64	3.05	.64
STRIKEOUTS	Bob Feller	1946	Cleveland Indians	348	113.37	67.45
WHIP	Tiny Bonham	1942	New York Yankees	0.99	1.34	.14
WINS	James Bagby	1920	Cleveland Indians	31	14.88	6.37

Expansion Era (1961-1975)

Baseball's most growth-oriented era happened between 1961 and 1975. The Expansion Era was marked by an explosion of new teams and new trends. Right before the Expansion Era began, there were some significant team moves. In 1953, the Boston Braves moved to Milwaukee; two years later, the Athletics of Philadelphia fled to Kansas City. Similarly, at the close of the 1957 campaign, the Dodgers and Giants broke the hearts of fans in Brooklyn and Manhattan, respectively, as they "caught the last train for the coast,[27]" and headed west for Los Angeles and San Francisco. And at the end of the 1960 season, the hapless Washington Senators left for Minnesota (now, the Minnesota Twins) in search of greener pastures. These moves set the stage for the birth of new franchises in the Expansion Era as Major League Baseball decided to add two new teams to each league. In 1961, the American League added a new Washington Senators team (which became the Texas Rangers in 1975), and Los Angeles welcomed the Angels. In 1962, the National League added the New York Mets and Houston Colt .45s, whose name was changed to the Astros in 1965. Meanwhile, the Braves left Milwaukee for Atlanta in 1966, and the Athletics fled Kansas City to Oakland in 1968. In 1969, Major League Baseball expanded again, with franchises awarded to Kansas City (the Royals), Seattle (the Pilots), San Diego (the Padres) and Montreal (the Expos). While the Royals, Padres and Expos were able to build lasting legacies in their respective cities (though the Expos moved to Washington in 2005, and became the Nationals) the Pilots quickly moved to Milwaukee (replacing the recently departed Braves) after just one season and changed their name to the Brewers.

With more teams in the league, and more players needed to fill the rosters of those teams, it is often argued that in this era the Major League Baseball talent pool started to become watered down, thus making it easier for elite players to put up even more impressive

---

[27] Lyrics from Don McLean's hit song "American Pie."

statistics as they outpaced an increasing number of peers playing long after their prime, or players who would never have made the major leagues in previous eras. It was during this 14-year period that pitchers became supermen, dominating at a level unforeseen since the game's infancy. Hurlers like Sandy Koufax, Bob Gibson and Tom Seaver became household names, capturing national attention as they picked apart opposing lineups.

While the Expansion Era certainly made it possible for more mediocre talent to enter the league (and anyone who bore witness to the folly that was the 1962 Mets would certainly agree), the demographic and sociocultural changes that were initiated in the Home Run Era came to full maturity, leading to increasing numbers of elite players as well. With racial and ethnic barriers now completely removed, there was an infusion of Black and Hispanic stars in the league – players who, in previous eras, were forced to play in the Negro Leagues or the professional leagues of myriad Latin American countries. Baseball benefitted from this injection of high-end talent, which likely offset, if not exceeded, the influx of mediocre talent resulting from expansion. Hence, while baseball's talent pool expanded to include players like Roderick "Hot Rod" Kanehl of the '62 Mets (a player with more career errors in his three seasons in the league than doubles, triples, home runs or RBI), who probably wouldn't have made a major league roster in previous eras, it also expanded to include stars from the Negro and Latin American leagues such as Willie Mays and Juan Marichal.

Also during this era, one of the most significant rule changes in baseball history was made in 1973 by the American League – one that was not only controversial, but had a significant impact on both hitting *and* pitching statistics. In its continuing effort to bring more offense to the game, Major League Baseball allowed for the use of a hitter who could have a regular slot in the batting order, but did not have to play the field. This "hitter," of course, would come to be known as the "Designated Hitter" (or DH for short). For the first time ever, a player could focus only on hitting, without having to play a position. This changed the game in so many ways. It allowed players to be hitting specialists without having to have major league-caliber fielding skills (either because their skills had diminished with age, or because they were always poor fielders). It also meant that pitchers now had to face nine impressive major-league hitters in each lineup, rather than eight hitters and a pitcher (who usually couldn't hit very well).

The DH was implemented only in the younger American League, and to this day the National League still does not use the DH. This difference has a direct and significant impact on statistics. It is also a common belief among many baseball players, observers and historians that until 2000 (when umpires worked games in both leagues), umpires in each league had different strike zones[28]. Though no official changes were made in the rule book regarding the strike zone in the Expansion Era, most people believed the strike zone to be bigger in the National League. A bigger strike zone could have an impact on many statistical categories – it could lead to more strikeouts, fewer walks and higher batting averages and on-base percentages (with hitters being forced to put more balls into play).

By 1968, the pitcher had grown into a machine – a manipulative craftsman gifted with finesse. In fact, the Don Drysdales, Juan Marichals and Jim Palmers of the world had become so dominant that the Rules Committee made changes to improve the offense's chance of getting a hit. The mound was lowered to 10 inches[29], the strike zone was made smaller and umpires were told to take a harder stance against illegal pitches. Some pitchers also claimed the ball's exterior was beginning to feel "harder" – a notable difference from the "softer" ball that may previously

---

[28] This was particularly the case in the years between 1973 (when the DH was implemented), and 2000 (when all umpires began working in both leagues).

[29] Grounds crews, however, have allegedly tailored their home mounds to benefit their pitching staffs – not fully abiding by the 10-inch rule (with Dodger Stadium, in particular, being rumored to be unusually high), and modifying the mound's slope.

have put hitters at a disadvantage. At season's end, Denny McClain and Bob Gibson won *both* the MVP *and* Cy Young awards for their respective leagues, the first and only time pitchers swept all four prizes in a single year.

Amid the expansion and tumult of this era, schedules were expanded to 162 games, and each league split into eastern and western divisions, leading to a revised playoff format. Needless to say, these changes (especially the expansion of the schedule from 154 to 162 games) had a meaningful effect on the raw statistics of this era, particularly for cumulative statistics such as home runs, RBI, strikeouts and wins.

TABLE 2.4A: CATEGORY LEADERS IN THE EXPANSION ERA (1961-1975)

HITTING STATISTICS

NATIONAL LEAGUE

STATISTIC	PLAYER	YEAR	TEAM	RAW VALUE	LEAGUE MEAN	LEAGUE SD
BATTING AVERAGE	Rico Carty	1970	Atlanta Braves	.366	2.83	.03
HOME RUNS	Willie Mays	1965	San Francisco Giants	52	16.86	12.03
OPS	Willie McCovey	1969	San Francisco Giants	1.108	.768	.11
RBI	Tommy Davis	1962	Los Angeles Dodgers	153	76.60	28.51

AMERICAN LEAGUE

STATISTIC	PLAYER	YEAR	TEAM	RAW VALUE	LEAGUE MEAN	LEAGUE SD
BATTING AVERAGE	Rod Carew	1974	Minnesota Twins	.364	.270	.03
HOME RUNS	Roger Maris	1961	New York Yankees	61	18.31	14.23
OPS	Norm Cash	1961	Detroit Tigers	1.148	.800	.13
RBI	James Gentile (tied)	1961	Baltimore Orioles	141	74.92	28.51
	Roger Maris (tied)	1961	New York Yankees			

TABLE 2.4B: CATEGORY LEADERS IN THE EXPANSION ERA (1961-1975)

PITCHING STATISTICS

NATIONAL LEAGUE

STATISTIC	PLAYER	YEAR	TEAM	RAW VALUE	LEAGUE MEAN	LEAGUE SD
ERA	Bob Gibson	1968	St. Louis Cardinals	1.12	2.86	.63
STRIKEOUTS	Sandy Koufax	1965	Los Angeles Dodgers	382	159.91	71.26
WHIP	Bob Gibson	1968	St. Louis Cardinals	0.85	1.17	.13
WINS	Steve Carlton	1972	Philadelphia Phillies	27	12.91	4.71

AMERICAN LEAGUE

STATISTIC	PLAYER	YEAR	TEAM	RAW VALUE	LEAGUE MEAN	LEAGUE SD
ERA	Luis Tiant	1968	Cleveland Indians	1.60	2.79	.57
STRIKEOUTS	Nolan Ryan	1973	California Angels	383	141.82	67.74
WHIP	Dave McNally	1968	Baltimore Orioles	0.84	1.13	.11
WINS	Denny McLain	1968	Detroit Tigers	31	13.91	4.80

Free Agency Era (1976-1985)

Although free agency did not formally begin until 1976, the winds of change began to blow seven years earlier. After the 1969 season, Cardinals' outfielder Curt Flood refused a trade to the Phillies, citing his desire to hear offers from other clubs before being sent off against his will. In doing so, he became the game's first unofficial free agent, and his efforts were validated when the Major League Baseball Players Association (MLBPA) – a.k.a., the player's union – backed his $4.1 million lawsuit against Major League Baseball.

In 1975, an arbitration panel heard the cases of Dave McNally and Andy Messersmith, both of whom opted for free agency after refusing to sign their contracts. In the end, it was ruled they could play for the highest bidder, prompting American League President Lee MacPhail to remark that players "might have to be paid more than they are worth."[30] Either way, a new precedent had been set.

In the summer of 1976, the owners and players agreed on a bargaining deal that included a free agent reentry draft, to be held in November. Players would be eligible for the draft after five seasons, at which point team owners could negotiate with as many free agents as they wanted. The first draft was held in New York, setting the stage for a slew of signings that off-season and beyond.

But there was trouble on the horizon: the frequent exchange of players raised concerns over roster replenishment. In early 1981, the owners announced free agent compensation rules that allotted an amateur draft pick to teams losing players in free agency. Unhappy with this new development (as compensation was perceived as a punishment for teams signing free agent players to large contracts), the MLBPA initiated a labor strike that summer when talks failed and federal mediators stepped in. From June 12 - August 10, 1981, there was no Major League Baseball. The impasse actually ended on

---

[30] Solomon, Burt (2001). *The Baseball Timeline*. DK Publishing, Inc.; New York.

July 31, but no regular season games were scheduled until after the All-Star Game on August 9.

At the strike's conclusion, the owners and MLBPA agreed to a pool compensation system that gave an amateur draft pick and an extra player to teams losing ranking free agents. The collective bargaining agreement was also lengthened under the deal. As for the damage, a total of 713 games were cancelled that season. The players lost an estimated $28 million in salary while the owners took a net loss of more than $70 million.

When Major League Baseball came back into play, there was a marked, and irreversible, financial trend upward, with player salaries and ticket prices on the rise. For example, in 1976, the average major league salary stood at $51,000; by 1980, it had risen to $143,000; and by 1981, $185,000 – more than triple what it had been just five years earlier.

In addition to the changes that may have resulted from free agency, it should be noted that this era was the first <u>full</u> era in which the DH was implemented. Furthermore, this era saw an innovation that, while not pioneered by Tony LaRussa, was refined and perfected by him: the use of late-inning, relief-pitching specialists. In previous eras, the pitchers who were relegated to the bullpen were often those considered less talented than the team's starters and only entered the game when it was out-of-hand. LaRussa (first with the Chicago White Sox, then the Oakland A's and eventually the St. Louis Cardinals) refined the practice of mixing and matching pitchers based on hitters' strengths and weaknesses. Most often, this meant matching a left-handed relief pitcher against a left-handed hitter (or a right-handed pitcher against a right-handed hitter) late in games because it is harder for hitters to hit against pitchers who throw from the same side as they hit.

This change, along with the more widespread use of 9th inning closers (like Rich "Goose" Gossage) and 8th inning setup men (like Ron Davis, who pioneered the role, setting up for Goose Gossage on the Yankees teams of the late '70s and early '80s), had a significant impact in a number of areas. First, it meant that starting pitchers didn't have to pitch as deep into games anymore. Starting pitchers could be given more rest for their next start, rather than spending more time in a game pitching when they were tired or ineffective. In some cases, the increased use of relief pitchers also helped starting pitchers get more wins because relief pitchers could take over for them while their team held the lead. However, in other cases, the increased use of relief pitchers could work against starting pitchers' statistics – for example, if they were removed from a game before their team took a late-inning lead, the win would not be awarded to them.

Similarly, the increased use of relief pitching specialists could both help and hurt starting pitchers' ERA and WHIP statistics, depending on the situation. In one case, if starters pitched fewer innings when they were tired, this could result in them giving up fewer earned runs, walks and hits; however, pitching fewer innings has a negative impact on both ERA and WHIP when all other things are kept equal. When it comes to cumulative pitching statistics such as strikeouts, the fewer innings pitched by starters (as a result of increased use of relief pitchers) would translate into lower totals for these statistical categories.

On the offensive side, the increased use of relief pitching specialists would have an inverse impact for hitters. Increased use of relief-pitching specialists means that, generally, hitters now face a pitcher late in games who is likely more effective than a tired starter and also likely throws from the same side in which the hitter hits, which is more challenging for most hitters. Though the full impact of the major innovation of this era – relief pitching specialists – is not completely clear, what is clear is that this is the era in which relief pitchers finally began to earn respect. Before this era, no relief pitcher had ever won the MVP Award, and only one had ever won a Cy Young Award (and that was just two years prior to the start of this era when Mike Marshall won the award in 1974). In this era alone, four relief pitchers won the Cy Young Award (Sparky Lyle,

Bruce Sutter, Rollie Fingers and Willie Hernandez) and two of them won the MVP Award in the same season (Rollie Fingers and Willie Hernandez). Since then, four other relief pitchers have won the Cy Young (Steve Bedrosian, Mark Davis, Dennis Eckersly and Eric Gagne), with one of them (Eckersly) also winning the MVP.

TABLE 2.5A: CATEGORY LEADERS IN THE FREE AGENCY ERA (1976-1985)

HITTING STATISTICS

NATIONAL LEAGUE

STATISTIC	PLAYER	YEAR	TEAM	RAW VALUE	LEAGUE MEAN	LEAGUE SD
BATTING AVERAGE	Willie McGee	1985	St. Louis Cardinals	.353	.271	.03
HOME RUNS	Dave Kingman	1979	Chicago Cubs	48	14.71	11.06
OPS	Mike Schmidt	1981	Philadelphia Phillies	1.08	.734	.10
RBI	George Foster	1977	Cincinnati Reds	149	70.76	25.56

AMERICAN LEAGUE

STATISTIC	PLAYER	YEAR	TEAM	RAW VALUE	LEAGUE MEAN	LEAGUE SD
BATTING AVERAGE	George Brett	1980	Kansas City Royals	.390	.287	.03
HOME RUNS	Jim Rice	1978	Boston Red Sox	46	14.32	10.17
OPS	George Brett	1980	Kansas City Royals	1.118	.782	.09
RBI	Don Mattingly	1985	New York Yankees	145	75.44	23.72

TABLE 2.5B: CATEGORY LEADERS IN THE FREE AGENCY ERA (1976-1985)

PITCHING STATISTICS

NATIONAL LEAGUE

STATISTIC	PLAYER	YEAR	TEAM	RAW VALUE	LEAGUE MEAN	LEAGUE SD
ERA	Dwight Gooden	1985	New York Mets	1.53	3.24	.67
STRIKEOUTS	J.R. Richard	1979	Houston Astros	313	120.18	46.75
WHIP	John Tudor	1985	St. Louis Cardinals	0.94	1.23	.15
WINS	Dwight Gooden	1985	New York Mets	24	13.45	4.93

AMERICAN LEAGUE

STATISTIC	PLAYER	YEAR	TEAM	RAW VALUE	LEAGUE MEAN	LEAGUE SD
ERA	Ron Guidry	1978	New York Yankees	1.74	3.60	.68
STRIKEOUTS	Nolan Ryan	1977	California Angels	341	125.98	53.41
WHIP	Ron Guidry	1978	New York Yankees	0.95	1.30	.13
WINS	Ron Guidry	1978	New York Yankees	25	13.40	4.55

Steroid Era (1986-2007)

Since the game's inception, baseball players have always tried mightily to gain an advantage over their competition. Some pitchers were notorious for using doctored balls, and some hitters were known to use corked bats. Both pitchers *and* hitters have also tried to get an edge by stealing signs. In more recent decades, steroids have taken over as the predominant method of cheating, giving players – often hitters – a stimulating edge. Whether it be cocaine and amphetamines, which were used most frequently in the 1970s and '80s, or anabolic steroids and human growth hormone that became ubiquitous in years since, performance-enhancing drugs (PEDs) are no stranger to our national pastime. This leads us to the Steroid Era: a two-decade span marked by federal intervention and some of the most prolific offense in baseball history.

The era loosely began in 1986, as this was the first full season that Jose Canseco played in the major leagues, and it was also the year after the initiation of baseball's Joint Drug Prevention and Treatment Program, initiated by the Commissioner of Baseball and the MLBPA. Jose Canseco, along with his fellow "Bash Brother" in Oakland, Mark McGwire (who debuted in 1986), would come to symbolize the Steroid Era. While Canseco is considered only a small part of the steroids story, his career helped put PED use into the national spotlight. After he retired, he wrote an exposé on the subject that had a profound impact on the national conversation about drug use in baseball[31].

In 1988, Canseco became the first player to hit 40 home runs and steal 40 bases in a single season. This was a year after McGwire won the AL Rookie of the Year Award, and two years after Canseco had won the same award himself. But Congress was cracking down on steroids, and in 1988, they passed the Anti-Drug Abuse Act, which criminalized the non-medicinal distribution or possession of

---

[31] *Juiced: Wild Times, Rampant 'Roids, Smash Hits & How Baseball Got Big*, Regan Books (2005).

anabolic steroids. Two years later, they passed the Anabolic Steroids Control Act, which placed steroids in an elevated drug class. In 1991, MLB Commissioner Fay Vincent added steroids to baseball's list of banned substances. But, while steroids may have been banned, they did not go away. There was no testing plan in place, and there wouldn't be for another decade.

Another players' strike in 1994 (no games were played from August 1994 to April 1995) deprived fans of more than 900 regular-season games, including the 1994 playoffs and World Series. When baseball finally returned, fans were angry, ambivalent and apathetic. By Memorial Day, 1995, attendance had diminished 21.5% from the year before. With attendance down and interest in baseball flagging, the game needed a jolt, and it came in the form of the Home Run Race of 1998. Mark McGwire and Sammy Sosa chased Roger Maris' sacrosanct, single-season home run record of 61, which was set all the way back in 1961. The race between McGwire and Sosa gave fans a reason to pay attention again and when the dust settled, both McGwire and Sosa broke Maris' record, with McGwire finishing on top with 70 dingers. As we all know, McGwire's new record didn't last long. Barry Bonds eclipsed the 70-homer mark just weeks after the catastrophic attacks of September 11, 2001.

With Bonds breaking McGwire's nascent record, Roger Maris' once-record-breaking total of 61, which had stood for nearly 40 years, had been surpassed for the third time in just four seasons. Fans had to wonder: was this just a coincidence? Or had something significant changed in baseball, making home runs much easier to hit?

It was neither a coincidence, nor the result of "something" having changed. It was the result of *many* things having changed in baseball during this era. The Steroid Era saw the construction of 19 new ballparks[32] – most of which were much friendlier to power

---

[32] **Comiskey Park II** in Chicago, IL (opened in 1991); **Oriole Park at Camden Yards** in Baltimore (opened in 1992); **Joe Robbie Stadium** in Miami (opened for baseball

hitters than the parks they replaced. In addition, four more expansion teams were added to MLB (the Arizona Diamondbacks; the Colorado Rockies; the Florida Marlins – now known as the Miami Marlins; and the Tampa Bay Devil Rays – now known as the Tampa Bay Rays). Both the AL and NL switched to a more tightly wound, "juiced" ball, according to the speculation of players and media analysts. Oh, and, there also seemed to be a dramatic increase in the number of players using steroids, along with a tendency by owners to *look the other way*, as documented by Canseco's book, and the 2007 bestseller, *Game of Shadows: Barry Bonds, BALCO, and the Steroids Scandal that Rocked Professional Sports* by San Francisco Chronicle journalists Mark Fainaru-Wada and Lance Williams. Put all of these things together, and what do you get? An *explosion* of home runs and other power statistics.

By the end of the 21st century's first decade, there was undeniable evidence that the Steroid Era's three offensive heroes – Barry Bonds, Mark McGwire and Sammy Sosa – had used steroids during their careers, and likely during their record-breaking seasons. In 2003, Bonds admitted[33] to using the designer steroids, "the cream and "the clear," created by the Bay Area Laboratory Co-operative (BALCO) earlier that season, but without knowing that they were

---

in 1993); **Rangers Ballpark in Arlington**, Arlington TX (opened in 1994); **Jacob's Field** in Cleveland (opened in 1994; now called **Progressive Field**); **Coors Field** in Colorado (opened in 1995); **Turner Field** in Atlanta, GA (opened in 1996); **Bank One Ballpark** in Arizona (opened in 1998, now called **Chase Field**); **Tropicana Field** in St. Petersburg, FL (opened for baseball in 1998); **Safeco Field** in Seattle, WA (opened in 1999); **Pacific Bell Park** in San Francisco (opened in 2000; now called **AT&T Park**); **Enron Field** in Houston (opened in 2000; now called **Minute Maid Park**); **Miller Park** in Milwaukee (opened in 2001); **PNC Park** in Pittsburgh (opened in 2001); **Comerica Park** in Detroit (opened in 2000); **Great American Ballpark** in Cincinnati (opened in 2003); **Citizens Bank Park** in Philadelphia (opened in 2004); **Petco Park** in San Diego (opened in 2004); **"New" Busch Stadium** in St. Louis (opened in 2006).

[33] Bonds made the admission during his 2003 grand jury testimony for the investigation of his personal trainer, Greg Anderson, as reported by Lance Williams & Mark Fainaru-Wada in an article published by the San Francisco Chronicle on December 12, 2004.

steroids. In his admission he said that he believed these substances were flaxseed oil and a rubbing balm for arthritis: treatments he received from trainer Greg Anderson. Subsequently, in 2007 Fainaru-Wada and Williams released *Game of Shadows*, which published incriminating leaked testimony from Troy Ellerman, a former attorney for BALCO founder Victor Conte, during the BALCO grand jury proceedings. Ellerman's testimony was supported by numerous documents and records alleging that Bonds used various steroids, including human growth hormone (HGH), during many years of his career, including the 2001 season when he broke McGwire's home run record. A few years later, in January of 2010, Mark McGwire finally came clean, admitting to using steroids for much of his Hall of Fame-caliber career, including during his 1998 record-breaking season. To this day, however, Sammy Sosa has not admitted to using steroids, though a 2009 investigative report from the New York Times[34] alleges that Sosa's name appeared on a 2003 list of players testing positive for a banned substance, but the substance he used was not specified. Oh, how the mighty have fallen, and with them, the records they set are now shrouded in controversy as the court of public opinion continues to deliberate on the merits of these records.

In addition to the direct effect of steroids during this period, the Steroid Era also absorbed the full impact of changes from previous eras, which picked up great steam at this time. For example, MLB experienced full integration of Black, Hispanic and now Asian players, from all over the world. In addition, during this era baseball ushered in a generation of hitters who spent their entire careers as DHs, like Edgar Martinez and David Ortiz. There was also the explosion of relief pitching specialists. On top of all these changes, pitching rotations changed once again, with five-man

---

[34] Michael S. Schmidt (June 16, 2009). "Sosa is Said to Have Tested Positive in 2003". *The New York Times.*

rotations now becoming the norm and starters getting 20% fewer starts than their counterparts in the four-man rotations of previous eras, which meant that they had 20% fewer opportunities for wins and strikeouts. By comparison, pitchers in the Pre-Modern and Deadball eras had more *wins* in a single season than most pitchers in the Steroid and Post-Steroid eras have had *starts*!

TABLE 2.6A: CATEGORY LEADERS IN THE STEROID ERA (1986-2007)

HITTING STATISTICS

NATIONAL LEAGUE

STATISTIC	PLAYER	YEAR	TEAM	RAW VALUE	LEAGUE MEAN	LEAGUE SD
BATTING AVERAGE	Tony Gwynn	1994	San Diego Padres	.394	.284	.03
HOME RUNS	Barry Bonds	2001	San Francisco Giants	73	22.42	14.78
OPS	Barry Bonds	2004	San Francisco Giants	1.422	.824	.12
RBI	Sammy Sosa	2001	Chicago Cubs	160	82.27	30.81

AMERICAN LEAGUE

STATISTIC	PLAYER	YEAR	TEAM	RAW VALUE	LEAGUE MEAN	LEAGUE SD
BATTING AVERAGE	Nomar Garciaparra	2000	Boston Red Sox	.372	.286	.03
HOME RUNS	Alex Rodriguez	2002	Texas Rangers	57	21.51	10.92
OPS	Frank Thomas	1994	Chicago White Sox	1.217	.818	.12
RBI	Manny Ramirez	1999	Cleveland Indians	165	85.63	28.13

TABLE 2.6B: CATEGORY LEADERS IN THE STEROID ERA (1986-2007)

PITCHING STATISTICS

NATIONAL LEAGUE

STATISTIC	PLAYER	YEAR	TEAM	RAW VALUE	LEAGUE MEAN	LEAGUE SD
ERA	Greg Maddux	1995	Atlanta Braves	1.56	4.02	.96
STRIKEOUTS	Randy Johnson	2001	Arizona Diamondbacks	372	155.93	51.34
WHIP	Greg Maddux	1995	Atlanta Braves	0.81	1.29	.15
WINS	Randy Johnson	2002	Arizona Diamondbacks	24	12.84	3.77

AMERICAN LEAGUE

STATISTIC	PLAYER	YEAR	TEAM	RAW VALUE	LEAGUE MEAN	LEAGUE SD
ERA	Pedro Martinez	2000	Boston Red Sox	1.74	4.47	.73
STRIKEOUTS	Pedro Martinez	1999	Boston Red Sox	313	140.17	40.63
WHIP	Pedro Martinez	2000	Boston Red Sox	0.74	1.41	.16
WINS	Bob Welch	1990	Oakland Athletics	27	13.61	4.04

Post-Steroid Era (2008-2014)

In 2001, MLB implemented its first random drug-testing program, but solely for the minor leagues. It wasn't until 2003 that testing reached the major-league level, however, at that time, the only penalty for first-time offenders was mandatory counseling. Punishment remained weak through 2005, until the MLB Joint Drug Prevention and Treatment Program was adopted in the spring of 2006 by Commissioner Bud Selig. This program mandated that players be subject to one unannounced mandatory test each year, along with random tests for selective players during both the season and off-season, with suspensions ranging from 10 days (for first-time offenders) to one year (for fourth-time offenders), without pay.

Despite the expansion of testing and more punitive disciplinary measures, MLB knew that more needed to be done. The issue came to a head in December of 2007 when a federal investigation into the use of steroids in baseball, spearheaded by former Senate Majority Leader George Mitchell, yielded the highly anticipated: *Report to the Commissioner of Baseball of an Independent Investigation into the Illegal Use of Steroids and Other Performance Enhancing Substances by Players in Major League Baseball* (a.k.a., the "Mitchell Report").

The Mitchell Report tied nearly 100 then-current and former players to illegal drug use, including stars like Roger Clemens, Andy Pettitte, Miguel Tejada and Éric Gagné. As a result of the Mitchell Report, MLB further expanded testing to two unannounced tests per year for all players and additional random testing for selective players, with testing now including an inspection for more substances (and at present, including HGH). Penalties also stiffened, with first-time offenders now receiving a 50-game suspension; second-time offenders receiving a 100-game suspension; and third-time offenders receiving a lifetime ban from baseball.

In the wake of the Mitchell Report, baseball fans and critics felt MLB had reached a new low. Players revered for a decade for their super-power achievements were now considered frauds. The game needed a clean generation to weed out the old guard, and a resurgence started, spurred by new stars and even newer ballparks. But the aftertaste of illegal drug use still lingers. The Mitchell Report did not end steroid use in baseball. Over the past year, Alex Rodriguez, who in February 2009 had previously admitted to using steroids from 2001 to 2003, was suspended, after a plea deal, for the entire 2014 season for steroid-related activities, as a result of MLB's investigation into the Biogenesis Clinic.[35] A-Rod's unprecedented suspension, which was initially levied on August 5, 2013, came on the heels of Biogenesis-related suspensions for 2011 National League MVP Ryan Braun and 12 other players.[36] These suspensions make clear that while we are technically in the "Post-Steroid" Era of baseball history, the problem hasn't been completely eradicated.

Still, most baseball observers and analysts believe that the Mitchell Report put a significant dent into the pervasive use of steroids that was all-too-common in the previous era, and the reduction in offensive statistics observed in the first few years of this new era support this opinion. As compared to the Steroid Era, home run totals and batting averages have been on the decline, but strikeout tallies have been rising. In fact, the number of perfect games thrown during the entire Steroid Era (between 1986 and 2007) – six – is equal to the number tossed in just the first seven seasons of the Post-Steroid Era (from 2008 to 2014, when this book was

---

[35] Shortly after receiving his suspension, which was originally a 211-game suspension (including the final 49 games of the 2013 season and all 162 games of the 2014 season), Rodriguez appealed which, according to MLB's collective bargaining agreement with the MLBPA, allowed him to continue playing until his appeal was to be heard at the end of the season. The appeal hearing was presided over by arbitrator Fredric Horowitz, who ruled in January, 2014, that Rodriguez's suspension be reduced to include only the 162-game schedule of the 2014 season.

[36] The other players include: Antonio Bastardo, Francisco Cervelli, Jordany Valdespin, Jesus Montero, Cesar Puello, Sergio Escalona, Fernando Martinez, Fautino De Los Santos, Jordan Norberto, Nelson Cruz, Everth Cabrera, Jhonny Peralta.

written). In addition, attitudes among players have changed, with significant numbers in the MLBPA signaling a willingness to cooperate with MLB to remove steroids from the game. High profile players, like Max Scherzer,[37] have also advocated for the voiding of contracts for steroid users, while others, like Skip Schumaker,[38] have called for lifetime bans for first time offenders.

Despite the taint of the Steroid Era, professional baseball continues to have its dramas and stars. Thus far during the Post-Steroid Era, the pitcher is once again a dominant force. Hurlers such as Justin Verlander, Felix Hernandez and Clayton Kershaw are members of the league's elite, giving fans a new class of ballplayers who are more trusted, with respect to steroids, than the previous generation. Despite the drop in offense, attendance hasn't faltered and fans continue to spend at the box office. Even when the housing bubble burst and the economy tanked in 2008, attendance totals for that year were the second-highest of all time. Following the 2012 campaign, the Commissioner's office reported that for each consecutive season from 2004 to 2012, Major League Baseball's cumulative attendance (for the entire league) was higher than that for any other season prior to 2004. Stadium construction also helped this trend, with four new ballparks opening since 2009. The Mets, Yankees, Twins and Marlins all bid farewell to their previous homes, and several other teams implemented more fan-friendly changes to their current structures.

---

[37] George Sipple (July 23rd, 2013). "Scherzer: Void Braun's $113M Contract." *The USA Today*.
[38] Steve Dilbeck (July 23rd, 2013). "Skip Schumaker says Ryan Braun should be suspended for life." *Los Angeles Times*.

Table 2.7a: Category Leaders in the Post-Steroid Era (2008-2014)

Hitting Statistics

National League

Statistic	Player	Year	Team	Raw Value	League Mean	League SD
Batting Average	Chipper Jones	2008	Atlanta Braves	.364	.279	.03
Home Runs	Ryan Howard	2008	Philadelphia Phillies	48	20.63	10.62
OPS	Albert Pujols	2008	St. Louis Cardinals	1.114	.815	.09
RBI	Ryan Howard	2008	Philadelphia Phillies	146	78.48	23.91

American League

Statistic	Player	Year	Team	Raw Value	League Mean	League SD
Batting Average	Joe Mauer	2009	Minnesota Twins	.365	.283	.03
Home Runs	Jose Bautista	2010	Toronto Blue Jays	54	17.56	10.61
OPS	Miguel Cabrera	2013	Detroit Tigers	1.078	.771	.09
RBI	Miguel Cabrera	2012	Detroit Tigers	139	74.48	21.10

TABLE 2.7B: CATEGORY LEADERS IN THE POST-STEROID ERA (2008-2014)

PITCHING STATISTICS

NATIONAL LEAGUE

STATISTIC	PLAYER	YEAR	TEAM	RAW VALUE	LEAGUE MEAN	LEAGUE SD
ERA	Clayton Kershaw	2014	Los Angeles Dodgers	1.77	3.43	.70
STRIKEOUTS	Tim Lincecum	2008	San Francisco Giants	268	149.64	39.07
WHIP	Clayton Kershaw	2014	Los Angeles Dodgers	0.86	1.23	.13
WINS	Brandon Webb	2008	Arizona Diamondbacks	22	12.24	3.69

AMERICAN LEAGUE

STATISTIC	PLAYER	YEAR	TEAM	RAW VALUE	LEAGUE MEAN	LEAGUE SD
ERA	Felix Hernandez	2014	Seattle Mariners	2.14	3.60	.76
STRIKEOUTS	Yu Darvish	2013	Texas Rangers	277	165.78	39.94
WHIP	Justin Verlander (tied)	2011	Detroit Tigers	0.92	1.24	.14
WHIP	Felix Hernandez (tied)	2014	Seattle Mariners	0.92	1.23	.13
WINS	Justin Verlander	2011	Detroit Tigers	24	12.64	3.50

CHAPTER 3

PRIMETIME LINEUP: BASEBALL'S Z-SCORE LEADERS

"Sometimes there's a man, well he's the man for his time and place."

- The Stranger,
The Big Lebowski

Category Leaders by Z-score, Across All Eras

Having identified the giants of each era according to the *raw* statistics of each player's single-season performance for the categories of interest, we can now report the z-scores for those top performances based on the league means and standard deviations for each season, in each category. In reporting these z-scores, we first present the lists of leaders for each respective statistical category of interest, as ranked by z-score. We then provide more general tables of our top 20 pitching z-scores, top 20 hitting z-scores and top 20 overall baseball z-scores, thus identifying the best of the best. Although the names on these lists will not surprise you, you may be surprised to learn which specific single-season performances top the charts. Hence, as you explore this chapter and immerse yourself in what the B-52s might call *Channel Z-score*, try to look beyond the face value of raw statistics – of strikeouts and homers, ERAs and OBPs – and allow the z-scores to spur your imagination as to why the single-season performances of the chart toppers were so special. If you get stumped, don't worry: we offer a cheat sheet at the end of the chapter.

Batting Average

Batting average, which is the simple ratio of hits-to-at-bats, is a traditional statistic with which most people are familiar. For much of baseball history, batting average was the statistic par excellence for determining the best hitters in the game. However, in present-day baseball, particularly among sabermetricians, batting average has fallen a bit out of favor, relative to OBP (on-base percentage), wOBA (weighted on-base average) and OPS (on-base plus slugging percentage). The main reason is because statistics like OBP, wOBA and OPS have been shown to be better predictors of team run production than batting average. Still, even the most pedestrian of sports fans is very familiar with the general concept of batting average as well as the common benchmarks for success (i.e., a .250

batting average will keep you in the big leagues; a career .300 batting average will get you into the Hall of Fame; and a .400 batting average during any single season will provide you with baseball immortality). As such, we included batting average in our selection of statistics. In Table 3.1, you will find the top batting averages of our analyses, across leagues and eras, ranked by z-score. In reviewing the table, you may find it curious that the man with the highest batting average in baseball history, Hugh Duffy (who posted a .440 average in 1894 for the Boston Braves), ranks all the way down at 11th on our list, while other players with lower batting averages (like Rod Carew, whose 1974 average of .364 was nearly 80 points lower than Duffy's) rank much higher. Why might this be the case? Though we discuss the reasons in greater detail at the end of the chapter, a quick look at the league mean for each of those seasons should give you a clue.

TABLE 3.1: Z-SCORE LEADERS FOR BATTING AVERAGE

RANK	PLAYER	TEAM	YEAR / LEAGUE (ERA)	RAW VALUE	LEAGUE MEAN	LEAGUE SD	Z-SCORE
1	Nap LaJoie	Philadelphia Athletics	1901 / AL (Deadball)	.426	.294	0.03	3.95
2	George Brett	Kansas City Royals	1980 / AL (Free Agency)	.390	.287	0.03	3.73
3	Tony Gwynn	San Diego Padres	1994 / NL (Steroid)	.394	.284	0.03	3.45
4	Rod Carew	Minnesota Twins	1974 / AL (Expansion)	.364	.270	0.03	3.39
5	Chipper Jones	Atlanta Braves	2008 / NL (Post-Steroid)	.364	.279	0.03	3.32
6	George Sisler	St. Louis Browns	1922 / AL (Home Run)	.420	.303	0.04	3.26
7	Rogers Hornsby	St. Louis Cardinals	1924 / NL (Home Run)	.424	.303	0.04	3.22
8	Willie McGee	St. Louis Cardinals	1985 / NL (Free Agency)	.353	.271	0.03	3.06
9	Joe Mauer	Minnesota Twins	2009 / AL (Post-Steroid)	.365	.283	0.03	2.98
10	Cy Seymour	Cincinnati Reds	1905 / NL (Deadball)	.377	.277	0.03	2.95
11	Hugh Duffy	Boston Braves	1894 / NL (Pre-Modern)	.440	.320	0.04	2.90
12	Nomar Garciaparra	Boston Red Sox	2000 / AL (Steroid)	.372	.286	0.03	2.83
13	Rico Carty	Atlanta Braves	1970 / NL (Expansion)	.366	.283	0.03	2.77

Home Runs

"Chicks dig the long ball." Those were the words of Tom Glavine, speaking to fellow Braves hurler Greg Maddux, in a famous Nike commercial from the late 1990s as they jealously observed the attention Mark McGwire was getting from some female fans. Fans have always been attracted to home runs. The power of the home run to attract attention has been understood as early as the 1920s when Babe Ruth's prowess to hit the long ball led to significant attendance increases, and a New York baseball team being able to finance a new stadium: a venue officially named Yankee Stadium, but known to everyone else as *The House That Ruth Built*. After the 1994 players strike, baseball had suffered a spiritual blow that contributed to a drop in attendance and interest in the game. However, it was the home run that once again rescued the sport as the Home Run Chase of 1998 between Mark McGwire and Sammy Sosa, who both broke Roger Maris' record of 61 home runs that season, resuscitated attendance figures and interest in the sport. The Home Run Chase of 1998 was a bittersweet phenomenon for baseball, because, although it brought renewed energy to the game, it came at a price: increased attention to the rampant use of steroids among the game's biggest stars, including Sosa, McGwire and Barry Bonds, and the complicity of owners who looked the other way in the interest of greater profits.

As noted throughout the book, an analysis of home runs using z-scores, particularly between the present-day single-season home run champion, Barry Bonds, and the champions of prior eras, before steroids were used, like Roger Maris and Babe Ruth, provided the impetus for this book. Which home run performances reign supreme when ranked by z-score? Take a look at Table 3.2 to find out.

TABLE 3.2: Z-SCORE LEADERS FOR HOME RUNS

RANK	PLAYER	TEAM	YEAR / LEAGUE (ERA)	RAW VALUE	LEAGUE MEAN	LEAGUE SD	Z-SCORE
1	Babe Ruth	New York Yankees	1919 / AL (Deadball)	29	3.47	4.48	5.70
2	Gavvy Cravath	Philadelphia Phillies	1915 / NL (Deadball)	24	3.82	4.21	4.79
3	Babe Ruth	New York Yankees	1927 / AL (Home Run)	60	6.91	11.28	4.71
4	Ned Williamson	Chicago Cubs	1884 / NL (Pre-Modern)	27	4.57	5.69	3.94
5	Jose Bautista	Toronto Blue Jays	2010 / AL (Post-Steroid)	54	17.56	10.61	3.43
6	Barry Bonds	San Francisco Giants	2001 / NL (Steroid)	73	22.42	14.78	3.42
7	Hack Wilson	Chicago Cubs	1930 / NL (Home Run)	56	14.14	12.42	3.37
8	Alex Rodriguez	Texas Rangers	2002 / AL (Steroid)	57	21.51	10.92	3.25
9	Jim Rice	Boston Red Sox	1978 / AL (Free Agency)	46	14.32	10.17	3.12
10	Dave Kingman	Chicago Cubs	1979 / NL (Free Agency)	48	14.71	11.06	3.01
11	Roger Maris	New York Yankees	1961 / AL (Expansion)	61	18.31	14.23	3.00
12	Willie Mays	San Francisco Giants	1965 / NL (Expansion)	52	16.86	12.03	2.92
13	Ryan Howard	Philadelphia Phillies	2008 / NL (Post-Steroid)	48	20.63	10.62	2.58

Runs Batted In (RBI)

RBI total is another traditional baseball statistic with which most casual fans are familiar. In recent years, however, RBI as a statistic, like batting average, has lost its luster, with sabermetricians favoring newer stats, like Runs Produced (Runs Scored plus RBI, minus Home Runs), instead. Nevertheless, with most fans having an implicit understanding of RBI as the traditional benchmark for success in a season (i.e., 100 RBI), we selected RBI as a statistical category of interest and list the top performers, ranked by z-score, in Table 3.3.

TABLE 3.3: Z-SCORE LEADERS FOR RBI

RANK	PLAYER	TEAM	YEAR / LEAGUE (ERA)	RAW VALUE	LEAGUE MEAN	LEAGUE SD	Z-SCORE
1	Sam Thompson	Detroit Wolverines	1887 / NL (Pre-Modern)	166	64.94	26.40	3.83
2	Gavvy Cravath	Philadelphia Phillies	1913 / NL (Deadball)	128	57.6	20.38	3.46
3	Miguel Cabrera	Detroit Tigers	2012 / AL (Post-Steroid)	139	74.48	21.10	3.15
4	Lou Gehrig	New York Yankees	1931 / AL (Home Run)	184	82.87	32.65	3.10
5	George Foster	Cincinnati Reds	1977 / NL (Free Agency)	149	70.76	25.56	3.07
6	Don Mattingly	New York Yankees	1985 / AL (Free Agency)	145	75.44	23.72	2.93
7	Hack Wilson	Chicago Cubs	1930 / NL (Home Run)	191	90.84	34.20	2.93
8	Manny Ramirez	Cleveland Indians	1999 / AL (Steroid Era)	165	85.63	2.82	2.82
9	Ryan Howard	Philadelphia Phillies	2008 / NL (Post-Steroid)	146	78.48	23.91	2.82
10	Tommy Davis	Los Angeles Dodgers	1962 / NL (Expansion)	153	76.6	28.51	2.68
11	Frank Baker	Philadelphia Athletics	1912 / AL (Deadball)	130	64.83	25.67	2.54
12	Sammy Sosa	Chicago Cubs	2001 / NL (Steroid)	160	82.27	30.81	2.52
13	James Gentile (tied)	Baltimore Orioles	1961 / AL (Expansion)	141	74.92	28.51	2.32
14	Roger Maris (tied)	New York Yankees	1961 / AL (Expansion)	141	74.92	28.51	2.32

OPS (On-Base Percentage plus Slugging)

Among the new sabermetric statistics, OPS seems to have caught on more quickly than many of the others. Perhaps the reason is because it is one of the easiest to compute, with it simply being the sum of a player's on-base percentage (OBP) and Slugging Percentage. Despite this, the benchmarks for success in this category are still a bit esoteric, with few outside of the cult of sabermetrics and hardcore baseball fans knowing what constitutes a good OPS figure. Average performance for OPS is usually regarded as being higher than about .700, and outstanding performance is often regarded as being higher than .900. As will be discussed in subsequent chapters, one of the advantages of z-scores is that, no matter what statistic you may be examining – a simple, traditional statistic of a single measurement, or a complex, new-fangled sabermetric statistic that combines a number of other esoteric statistics and weights – the range of z-scores will always be the same, with the mean being zero; good performance being considered a score of about +1.0 or better; and outstanding performance being considered a score of +2.0 or better. On the following page, in Table 3.4, you will find the top performers for OPS, ranked by z-score.

TABLE 3.4: Z-SCORE LEADERS FOR OPS

RANK	PLAYER	TEAM	YEAR / LEAGUE (ERA)	RAW VALUE	LEAGUE MEAN	LEAGUE SD	Z-SCORE
1	Barry Bonds	San Francisco Giants	2004 / NL (Steroid)	1.422	0.824	0.119	5.02
2	Babe Ruth	New York Yankees	1920 / AL (Home Run)	1.379	0.795	0.14	4.17
3	George Brett	Kansas City Royals	1980 / AL (Free Agency)	1.118	0.782	0.086	3.89
4	Rogers Hornsby	St. Louis Cardinals	1925 / NL (Home Run)	1.245	0.837	0.113	3.60
5	Babe Ruth	New York Yankees	1919 / AL (Deadball)	1.114	0.735	0.106	3.59
6	Miguel Cabrera	Detroit Tigers	2013 / AL (Post-Steroid)	1.078	0.771	0.087	3.54
7	Mike Schmidt	Philadelphia Phillies	1981 / NL (Free Agency)	1.080	0.734	0.099	3.49
8	Frank Thomas	Chicago White Sox	1994 / AL (Steroid)	1.217	0.818	0.117	3.41
9	Albert Pujols	St. Louis Cardinals	2008 / NL (Post-Steroid)	1.114	0.815	0.093	3.21
10	Willie McCovey	San Francisco Giants	1969 / NL (Expansion)	1.108	0.768	0.112	3.04
11	Heinie Zimmerman	Chicago Cubs	1912 / NL (Deadball)	0.989	0.761	0.079	2.90
12	Hugh Duffy	Boston Braves	1894 / NL Pre-Modern)	1.196	0.850	0.123	2.83
13	Norm Cash	Detroit Tigers	1961 / AL (Expansion)	1.148	0.800	0.125	2.82

Earned Run Average (ERA)

Just as batting average has long been used as the statistic par excellence to determine the pecking order of hitters, ERA has traditionally been used to identify the league's best pitchers. Like batting average, the ERA statistic requires the calculation of a simple formula:

$$\text{ERA} = \frac{Earned\ Runs\ Yielded}{Innings\ Pitched} \times 9\ (\text{innings})$$

At present, ERA is still regarded as an important statistic to evaluate pitching prowess, however, other sabermetric statistics, like WHIP (Walks plus Hits, divided by Innings Pitched) are now used more commonly to supplement the evaluation.

The year 1968 is often regarded as "the year of the pitcher," with both Bob Gibson and Denny McLain winning the MVP, and with Luis Tiant joining Gibson in posting a microscopic ERA below 2.00. Yet, none of these pitchers top our list of ERA leaders when ranked by z-score. Why? Once again, though a more detailed discussion of the reasons is provided at the end of the chapter, a quick look at the league mean for each of those seasons should give you a clue. In Table 3.5 we list the top performers for ERA, ranked by z-score.

TABLE 3.5: Z-SCORE LEADERS FOR ERA

RANK	PLAYER	TEAM	YEAR / LEAGUE (ERA)	RAW VALUE	LEAGUE MEAN	LEAGUE SD	CORRECTED Z-SCORE
1	Pedro Martinez	Boston Red Sox	2000 / AL (Steroid)	1.74	4.47	0.73	3.75
2	Dutch Leonard	Boston Red Sox	1914 / AL (Deadball)	0.96	2.65	0.56	3.00
3	Bob Gibson	St. Louis Cardinals	1968 / NL (Expansion)	1.12	2.86	0.63	2.76
4	Ron Guidry	New York Yankees	1978 / AL (Free Agency)	1.74	3.6	0.68	2.74
5	Mordecai Brown	Chicago Cubs	1906 / NL (Deadball)	1.04	2.51	0.55	2.68
6	Greg Maddux	Atlanta Braves	1994 / NL (Steroid)	1.56	4.02	0.96	2.58
7	Dwight Gooden	New York Mets	1985 / NL (Free Agency)	1.53	3.24	0.67	2.57
8	Carl Hubbell	New York Giants	1933 / NL (Home Run)	1.66	3.16	0.59	2.53
9	Clayton Kershaw	Los Angeles Dodgers	2014 / AL (Post-Steroid)	1.77	3.43	0.70	2.30
10	Spud Chandler	New York Yankees	1943 / AL (Home Run)	1.64	3.05	0.64	2.22
11	Luis Tiant	Cleveland Indians	1968 / AL (Expansion)	1.60	2.79	0.57	2.06
12	Felix Hernandez	Seattle Mariners	2014 / AL (Post-Steroid)	2.14	3.60	0.76	1.91
13	Tim Keefe	Troy Trojans	1880 / NL (Pre-Modern)	0.86	2.42	0.83	1.87

Wins

Long before Charlie Sheen made a fetish out of "winning," the concept was fairly popular. Ask legendary Green Bay Packers coach, Vince Lombardi, about the subject and he'd say: "Winning isn't everything, it's the only thing!" Similarly, Al Davis, the former owner of the Oakland Raiders, implored his players to "Just Win, Baby," by almost any means necessary.

For over a century within baseball, an evaluation of a pitcher's propensity to acquire "wins" was most commonly used to distinguish the best from the rest. In previous eras of baseball history, when there were fewer pitchers in each starting rotation, working longer into games and making many more starts than they do today, compiling a single-season win total over 30 was a relatively common occurrence. Today, a 30-win season is almost impossible, with five-man rotations leading to fewer starts than in previous eras and relief specialists routinely ending a starter's outing after six innings. The last man to win 30 games in a single season was Denny McLain in "the year of the pitcher," 1968, and it is likely that this threshold won't be passed again. Still, just how good of an indicator of pitching prowess can a statistic be that is highly dependent on a pitcher's *team's* ability to score runs? The answer to this question is debatable. Some sabermetricians believe that win totals are not to be taken seriously, while other analysts believe that this statistic still has merit, given that some pitchers *just know how to win*, regardless of other factors like run support, while others always seem to *pitch just well enough to lose*. Whichever side of the fence you are on, we believe that examining z-scores of win totals can help aid the evaluation, particularly across eras, controlling for factors like the size of pitching rotations and the height of pitching mounds, which have both changed through the years. In Table 3.6, we list the top performers for wins, ranked by z-score.

TABLE 3.6: Z-SCORE LEADERS FOR WINS

RANK	PLAYER	TEAM	YEAR / LEAGUE (ERA)	RAW VALUE	LEAGUE MEAN	LEAGUE SD	Z-SCORE
1	Denny McLain	Detroit Tigers	1968 / AL (Expansion)	31	13.91	4.8	3.56
2	Jack Chesbro	New York Highlanders	1904 / AL (Deadball)	41	16.03	7.3	3.42
3	Bob Welch	Oakland Athletics	1990 / AL (Steroid)	27	13.61	4.04	3.31
4	Justin Verlander	Detroit Tigers	2011 / AL (Post-Steroid)	24	12.64	3.5	3.24
5	Christy Matthewson	New York Giants	1908 / NL (Deadball)	37	14.56	7.03	3.19
6	Steve Carlton	Philadelphia Phillies	1972 / NL (Expansion)	27	12.91	4.71	2.99
7	Randy Johnson	Arizona Diamondbacks	2002 / NL (Steroid)	24	12.84	3.77	2.96
8	Dizzy Dean	St. Louis Cardinals	1934 / NL (Home Run)	30	15.24	5	2.95
9	Brandon Webb	Arizona Diamondbacks	2008 / NL (Post-Steroid)	22	12.24	3.69	2.65
10	Charley Radbourn	Providence Grays	1884 / NL (Pre-Modern)	59	17.54	16.15	2.57
11	Ron Guidry	New York Yankees	1978 / AL (Free Agency)	25	13.4	4.55	2.55
12	James Bagby	Cleveland Indians	1920 / AL (Home Run)	31	14.88	6.37	2.53
13	Dwight Gooden	New York Mets	1985 / NL (Free Agency)	24	13.45	4.93	2.15
14	Catfish Hunter	Oakland Athletics	1974 / AL (Free Agency)	25	14.93	5.37	1.88
15	Ferguson Jenkins	Texas Rangers	1974 / AL (Free Agency)	25	14.93	5.37	1.88

Strikeouts

On June 17, 1978, with the California Angels on their way to being struck out 18 times by the pitcher known as "Louisiana Lightning" at Yankee Stadium, a tradition was born. For the last few innings of the game, each time Yankees pitcher Ron Guidry had two strikes on the batter, the crowd rose to its feet in anticipation of a strikeout. Just as a home run is the most exciting offensive event in the game of baseball, a strikeout is the most exciting feat that a pitcher can achieve. It is also one of the few statistics that both old school and new school baseball analysts can agree on in terms of its predictive value in evaluating pitchers. The fewer balls put into play, the fewer scoring opportunities for an opposing team, and a strikeout is still the only way to record an out on a batter without him putting the ball in play.

To strike out a major league hitter is no easy feat. It requires a combination of control, *stuff* and deception. The pitchers in Table 3.7 have varying combinations of these three traits. Who tops the list? The answer may surprise you.

TABLE 3.7: Z-SCORE LEADERS FOR STRIKEOUTS

RANK	PLAYER	TEAM	YEAR / LEAGUE (ERA)	RAW VALUE	LEAGUE MEAN	LEAGUE SD	Z-SCORE
1	Dazzy Vance	Brooklyn Dodgers	1924 / NL (Home Run)	262	65.62	38.27	5.13
2	Pedro Martinez	Boston Red Sox	1999 / AL (Steroid)	313	140.17	40.63	4.25
3	Randy Johnson	Arizona Diamondbacks	2001 / NL (Steroid)	372	155.93	51.34	4.21
4	J.R. Richard	Houston Astros	1979 / NL (Free Agency)	313	120.18	46.75	4.12
5	Nolan Ryan	California Angels	1977 / AL (Home Run)	341	125.98	53.41	4.03
6	Rube Waddell	Philadelphia Athletics	1904 / AL (Deadball)	349	129.51	58.39	3.76
7	Nolan Ryan	California Angels	1973 / AL (Expansion)	383	141.82	67.74	3.73
8	Christy Matthewson	New York Giants	1903 / NL (Deadball)	267	101.1	44.83	3.70
9	Bob Feller	Cleveland Indians	1946 / AL (Home Run)	348	113.37	67.45	3.48
10	Sandy Koufax	Los Angeles Dodgers	1965 / NL (Expansion)	382	159.91	71.26	3.12
11	Tim Lincecum	San Francisco Giants	2008 / NL (Post-Steroid)	268	149.64	39.07	2.95
12	Yu Darvish	Texas Rangers	2013 / AL (Post-Steroid)	277	165.78	39.94	2.78
13	Charley Radbourn	Providence Grays	1884 / NL (Pre-Modern)	441	169.38	121.86	2.23

WHIP (Walks plus Hits, divided by Innings Pitched)

In the new economy of baseball statistics, the most highly valued currency is team runs, and the best predictors of team runs, according to sabermetricians, tend to be statistics that most directly relate to getting on base, regardless of how it is done. As such, the sabermetric statistic of WHIP has gained increasing favor in recent years because it evaluates pitching performance in a way that more accurately accounts for the allowance of total base runners. As noted above, the formula for WHIP is:

$$\text{WHIP} = \frac{Walks + Hits}{Innings\ Pitched}$$

As with the hitting statistic OPS, most casual fans are not familiar with the concept of WHIP, nor what constitutes a good WHIP coefficient. As noted with OPS, z-scores can help in this regard as they have the same evaluative properties, without regard to the measure being evaluated. In general, WHIP coefficients that are less than 1.0 are considered outstanding. The lowest WHIP coefficient of all time was for Pedro Martinez in 2000, at the height of the Steroid Era, when he posted a 0.74 WHIP for the Boston Red Sox. As you can imagine, this accomplishment, considering when it was achieved, led him to be our z-score leader in this category, as illustrated in Table 3.8.

TABLE 3.8: Z-SCORE LEADERS FOR WHIP

RANK	PLAYER	TEAM	YEAR / LEAGUE (ERA)	RAW VALUE	LEAGUE MEAN	LEAGUE SD	CORRECTED Z-SCORE
1	Pedro Martinez	Boston Red Sox	2000 / AL (Steroid)	0.74	1.41	0.16	4.20
2	Walter Johnson	Wash. Senators	1913 / AL (Deadball)	0.78	1.24	0.14	3.32
3	Greg Maddux	Atlanta Braves	1995 / NL (Steroid)	0.81	1.29	0.15	3.26
4	Clayton Kershaw	Los Angeles Dodgers	2014 / NL (Post-Steroid)	0.86	1.23	0.13	2.80
5	Ron Guidry	New York Yankees	1978 / AL (Free Agency)	0.95	1.30	0.13	2.69
6	Dave McNally	Baltimore Orioles	1968 / AL (Expansion)	0.84	1.13	0.11	2.52
7	Tiny Bonham	New York Yankees	1942 / AL (Home Run)	0.99	1.34	0.14	2.49
8	Bob Gibson	St. Louis Cardinals	1968 / NL (Expansion)	0.85	1.17	0.13	2.46
9	Felix Hernandez (tied)	Seattle Mariners	2014 / AL (Post-Steroid)	0.92	1.23	0.13	2.39
10	Christy Matthewson	New York Mets	1909 / NL (Deadball)	0.83	1.17	0.15	2.32
11	Justin Verlander (tied)	Detroit Tigers	2011 / AL (Post-Steroid)	0.92	1.24	0.14	2.28
12	Warren Hacker	Chicago Cubs	1952 / NL (Home Run)	0.95	1.30	0.15	2.26
13	John Tudor	St. Louis Cardinals	1985 / NL (Free Agency)	0.94	1.23	0.15	1.97
14	Tim Keefe	Troy Trojans	1880 / NL (Pre-Modern)	0.80	1.12	0.21	1.56

The Best of the Best

In the tables presented earlier in this chapter we listed the top performers in each specific statistical category, as ranked by z-score. In the tables presented on the subsequent pages we identify the *best of the best*, listing the top 20 single-season performances in the general categories of pitching and hitting, respectively, as well as the top 20 single-season performances for baseball as a whole.

TABLE 3.9: TOP 20 HITTING Z-SCORES

RANK	NAME	TEAM	YEAR	STAT	RAW VALUE	Z-SCORE
1	Babe Ruth	Boston Red Sox	1919	Home Runs	29	5.70
2	Barry Bonds	San Francisco Giants	2004	OPS	1.422	5.02
3	Gavvy Cravath	Philadelphia Phillies	1915	Home Runs	24	4.79
4	Babe Ruth	New York Yankees	1927	Home Runs	60	4.71
5	Babe Ruth	New York Yankees	1920	OPS	1.379	4.17
6	Nap Lajoie	Philadelphia Athletics	1901	Batting Average	.426	3.95
7	Ned Williamson	Chicago Cubs	1884	Home Runs	27	3.94
8	George Brett	Kansas City Royals	1980	OPS	1.118	3.89
9	Sam Thompson	Detroit Wolverines	1887	RBI	166	3.83
10	George Brett	Kansas City Royals	1980	Batting Average	.390	3.73
11	Rogers Hornsby	St. Louis Cardinals	1925	OPS	1.245	3.60
12	Babe Ruth	New York Yankees	1919	OPS	1.114	3.59
13	Miguel Cabrera	Detroit Tigers	2013	OPS	1.078	3.54
14	Mike Schmidt	Philadelphia Phillies	1981	OPS	1.080	3.49
15	Gavvy Cravath	Philadelphia Phillies	1913	RBI	128	3.46
16	Tony Gwynn	San Diego Padres	1994	Batting Average	.394	3.45
17	Jose Bautista	Toronto Blue Jays	2010	Home Runs	54	3.43
18	Barry Bonds	San Francisco Giants	2001	Home Runs	73	3.42
19	Frank Thomas	Chicago White Sox	1994	OPS	1.217	3.41
20	Rod Carew	Minnesota Twins	1974	Batting Average	.364	3.39

TABLE 3.10: TOP 20 PITCHING Z-SCORES

RANK	NAME	TEAM	YEAR	STAT	RAW VALUE	CORRECTED Z-SCORE
1	Dazzy Vance	Brooklyn Dodgers	1924	Strikeouts	262	5.13
2	Pedro Martinez	Boston Red Sox	1999	Strikeouts	313	4.25
3	Randy Johnson	Arizona Diamondbacks	2001	Strikeouts	372	4.21
4	Pedro Martinez	Boston Red Sox	2000	WHIP	0.74	4.20
5	J.R. Richard	Houston Astros	1979	Strikeouts	313	4.12
6	Nolan Ryan	California Angels	1977	Strikeouts	341	4.03
7	Rube Waddell	Philadelphia Athletics	1904	Strikeouts	349	3.76
8	Pedro Martinez	Boston Red Sox	2000	ERA	1.74	3.75
9	Nolan Ryan	California Angels	1973	Strikeouts	383	3.73
10	Christy Matthewson	New York Giants	1903	Strikeouts	267	3.70
11	Denny McLain	Detroit Tigers	1968	Wins	31	3.56
12	Bob Feller	Cleveland Indians	1946	Strikeouts	348	3.48
13	Jack Chesbro	New York Highlanders	1904	Wins	41	3.42
14	Walter Johnson	Washington Senators	1913	WHIP	0.78	3.32
15	Bob Welch	Oakland Athletics	1990	Wins	27	3.31
16	Greg Maddux	Atlanta Braves	1995	WHIP	0.81	3.26
17	Justin Verlander	Detroit Tigers	2011	Wins	24	3.24
18	Christy Matthewson	New York Giants	1908	Wins	37	3.19
19	Sandy Koufax	Los Angeles Dodgers	1965	Strikeouts	382	3.12
20	Dutch Leonard	Boston Red Sox	1914	ERA	0.96	3.00

TABLE 3.11: TOP 20 OVERALL Z-SCORES

RANK	NAME	TEAM	YEAR	STAT	RAW VALUE	Z-SCORE
1	Babe Ruth	Boston Red Sox	1919	Home Runs	29	5.70
2	Dazzy Vance	Brooklyn Dodgers	1924	Strikeouts	262	5.13
3	Barry Bonds	San Francisco Giants	2004	OPS	1.422	5.02
4	Gavvy Cravath	Philadelphia Phillies	1915	Home Runs	24	4.79
5	Babe Ruth	New York Yankees	1927	Home Runs	60	4.71
6	Pedro Martinez	Boston Red Sox	1999	Strikeouts	313	4.25
7	Randy Johnson	Arizona Diamondbacks	2001	Strikeouts	372	4.21
8	Pedro Martinez	Boston Red Sox	2000	WHIP	0.74	4.20
9	Babe Ruth	New York Yankees	1920	OPS	1.379	4.17
10	J.R. Richard	Houston Astros	1979	Strikeouts	313	4.12
11	Nolan Ryan	California Angels	1977	Strikeouts	341	4.03
12	Nap Lajoie	Philadelphia Athletics	1901	Batting Average	.426	3.95
13	Ned Williamson	Chicago Cubs	1884	Home Runs	27	3.94
14	George Brett	Kansas City Royals	1980	OPS	1.118	3.89
15	Sam Thompson	Detroit Wolverines	1887	RBI	166	3.83
16	Rube Waddell	Philadelphia Athletics	1904	Strikeouts	349	3.76
17	Pedro Martinez	Boston Red Sox	2000	ERA	1.74	3.75
18	Nolan Ryan	California Angels	1973	Strikeouts	383	3.73
19	George Brett	Kansas City Royals	1980	Batting Average	.390	3.73
20	Christy Matthewson	New York Giants	1903	Strikeouts	267	3.70

The Sultan of Swat: King of Kings

Admittedly, one of the main purposes of our investigation was to determine how Barry Bonds' record of 73 home runs, set in 2001, compared to the home run totals of previous record holders: particularly, Babe Ruth and Roger Maris. Though it may not come as a surprise that the z-score for Barry Bonds' 73 home runs (3.42) is considerably larger than Roger Maris' z-score (3.00) for the 61 home runs he hit in 1961, it *may* come as a surprise that Babe Ruth's z-score for his 60 home runs in 1927 dwarfs them both at 4.71. But even more surprising is that **the z-score for Ruth's 29 home runs in 1919 beats all of them at 5.70**! In fact, this figure not only represents the largest z-score we obtained for home runs, it is also the largest z-score we calculated for *any* baseball statistic.

Most people who follow baseball know that before Babe Ruth became the game's most prominent hitter with the New York Yankees, he was an outstanding pitcher for the Boston Red Sox. However, according to our z-score analyses, his most impressive *offensive* accomplishment – his z-score of 5.70 for hitting 29 home runs in 1919 – came during his final season with the *Boston Red Sox*. Perhaps if the Red Sox owner at the time, Harry Frazee, had understood just how impressive the Bambino's accomplishment was that season, he would never have agreed to the most infamous sports transaction of all time: selling Babe Ruth to the Yankees for $200,000 ($125,000 up front and $75,000 later) and a $300,000 loan, as legend has it, to pay for his Broadway show, *No, No, Nanette*.[39]

But how is it that Babe Ruth's 1919 home run total of 29 – which is less than half of his own highest total (of 60 home runs in 1927) and much less than Barry Bonds' record-setting total of 73 in 2001 – yielded the highest z-score for any baseball statistic in our analyses? The answer is that Babe Ruth's 1919 home run total was nearly 10 times the league average that season (which was 3.47). And since home runs were so uncommon in that era, the rest of the

---

[39] Robert W. Creamer (1974), *Babe: The Legend Comes to Life*.

players in the league were all very close to the league average, which led to there being a small standard deviation (4.48) for home runs. The combination of both of these factors helped Babe Ruth achieve a monumentally high z-score for home runs that season. As alluded to in Chapter 2, home runs in the Deadball Era were a rare occurrence for several reasons: the pitcher's mound could be as high as 15 inches; baseballs tended to be heavier or "dead;" and, for the most part, ballparks were absolutely enormous.

In contrast, when Barry Bonds set the existing home run record in 2001, his total of 73 was only about three times as large as the league average of 22.42, and the standard deviation was much larger (14.78) for that season. This was because there was a wider range of performance in the category of home runs that season (and for all seasons in the Steroid Era), likely influenced by the fact that some players were using performance-enhancing drugs (as Barry Bonds is believed to have done at the time) while others were not. Hence, while home runs were much easier to hit for ALL players in 2001, relative to 1919 – due to rampant steroid use, lowered pitcher's mounds, an allegedly "juiced" baseball, significantly smaller ballparks and diluted pitching staffs caused by league expansion and free agent compensation rules – these inflated home run totals were largely offset by the higher standard deviation for home runs (for the reasons discussed above). As such, *none* of the 24 players in 2001 who had more home runs than Babe Ruth's total of 29 in 1919 had a z-score that was nearly as big as Ruth's z-score of 5.70. Bonds' z-score of 3.42 was the closest, but is still *puny* in comparison.

To put Ruth's gargantuan z-score for home runs in 1919 into greater perspective, consider the following: If Babe Ruth had achieved the same z-score he had in 1919 (5.70) playing in the National League in 2001 (when Bonds hit 73 home runs), this would translate to hitting 107 home runs that season!⁴⁰,⁴¹ Conversely, if Bonds had achieved the same z-score he had in 2001 (3.42) playing in the American League in 1919, when Ruth hit 29 home runs, this would translate to a total of just 19 home runs.⁴² Though Bonds is best known for *playing* for the Giants, in this comparison it is clear that the Bambino is the *real* giant.

Dazzy Vance: The Z-score King of the Pitchers

While Babe Ruth is the undisputed z-score king – not only for hitting, but across all categories we examined – his z-score counterpart among pitchers is Dazzy Vance. In 1924, Dazzy Vance of the Brooklyn Dodgers recorded 262 strikeouts. That figure might not sound impressive today, when many of the game's top pitchers break the 300 strikeout threshold, but in 1924 strikeouts were much harder to come by. Though Vance's 1924 total of 262 strikeouts is currently ranked 130th on the list of single-season strikeout totals, this feat yielded the highest pitching z-score of our analyses: 5.13. Vance's z-score of 5.13 is nothing short of exceptional. It is the second highest z-score overall (across hitting and pitching categories) that we computed. Drawing from our example in Chapter 1 of how to use z-scores to compare height and IQ, Dazzy Vance's z-score is as remarkable as a man who stands seven feet tall and has an IQ of 177!

---

[40] In 2001, the National League mean for home runs was 22.42 and the standard deviation was 14.78. Hence, if we use algebra to figure out what Ruth's home run total would be (multiplying his Z-score of 5.70 by the standard deviation of 14.78, and then adding the mean, 22.42) we get 106.67, or 107 rounding up.

[41] Using the same formula, we determined that Roger Maris' Z-score of 3.00, which he achieved for hitting 61 home runs in 1961, would have translated into 67 home runs in the National League in 2001.

[42] Using the same formula, but with a mean of 3.47 and a standard deviation of 4.48, we determined that Barry Bonds' Z-score of 3.42, which he achieved for hitting 73 home runs in 2001, would have translated into 19 home runs in the American League in 1919.

Playing mostly in Brooklyn, Vance led the National League in strikeouts for seven straight seasons (1922-1928) and never had fewer than 130 in a single campaign. He tallied 2,045 strikeouts over the span of his 16-year career and in 1924 he was voted Most Valuable Player as he topped the league not only in strikeouts, but in wins and ERA as well. He was elected to the Hall of Fame in 1955.

How Great Was Pedro Martinez?

While Dazzy Vance might be the z-score king for a single season, in a single pitching category, Pedro Martinez's z-score greatness spans multiple seasons and multiple categories. In 1999 and 2000, Martinez achieved pitching feats for the Boston Red Sox that would be considered exceptional in any era. But when you consider that he put up those numbers during the heart of the Steroid Era – the most offensively dominant era in baseball history – it makes his achievements even more amazing. In 1999, Martinez's 313 strikeouts yielded a z-score of 4.25, which was the second-highest pitching z-score we computed and the sixth-highest overall z-score of our analyses. And his other statistics that season were almost equally impressive – 23 wins, 4 losses, 2.07 ERA, and 0.923 WHIP – even though they didn't yield z-scores that were on our top-20 lists.

In 2000, Martinez's 0.74 WHIP (which is still the lowest WHIP of all-time) yielded a z-score of 4.20, which is the fourth-highest z-score for pitching statistics and the eighth-highest overall z-score in our combined analyses. In addition, the 1.74 ERA he posted that season yielded a z-score of 3.75, which is the eighth-highest among pitching statistics and the 17th-highest overall in our analyses.

In sum, Pedro Martinez had three of the top eight z-scores that we calculated for pitchers, and three of the top 17 z-scores overall for pitchers and hitters combined. This leads us to conclude that Martinez – by virtue of his incredible seasons in 1999 and 2000 (which may be the best back-to-back seasons by a pitcher of all-time) and his entire body of work, which was within the most dominant offensive era in baseball history, the Steroid Era – was perhaps the greatest pitcher of all time. Care to argue? Go right ahead. But don't get too comfortable up there as Pedro is likely to knock you down with one of his "accidental" purpose pitches up-and-in.

Why Is Charley Radbourn's 59 Wins *NOT* Among the Top 20 Pitching Z-scores?

In 1884, Charley Radbourn had 59 wins for the Providence Grays. By today's standards, this is a mythic total. However, this amazing achievement yielded a z-score of just 2.57, which was not even close to making our top-20 list of pitching z-scores. By now, you may be able to hypothesize why this may be so, but if not, we will explain in greater detail below.

In 1884, one-to-two man pitching rotations led to star pitchers starting considerably more games than they do today (compare the range of starts – 31 to 73 – among team leaders in 1884, to those of 2012 – 31 to 33 starts). Radbourn had 73 starts in 1884. He started 65% of his team's games, which, at that time numbered around 112 games for most teams. High win totals were common in that era, and particularly in 1884, as five pitchers won at least 30 games and three pitchers won at least 40 games.

As for Charley Radbourn, he not only had a spectacular season in 1884, setting the record for wins which still stands, but he had a spectacular career, amassing 311 wins and nearly 500 complete games in 11 seasons. He led the National League in either strikeouts, wins, ERA or winning percentage in each of his first three seasons, and swept those categories in 1884. In 1939, Radbourn was voted into the Hall of Fame, along with Babe Ruth and the rest of the inaugural class. Radbourn played the bulk of his career with the Grays, based in Providence, Rhode Island. The Grays were added to the National League for the 1878 season and played through 1885. They were not part of the league's original slate of teams but ended up winning two titles during their abbreviated existence. The team folded after the 1885 season because of financial problems.

Nearly a century after Charley Radbourn's historic season, Denny McLain assembled one of the greatest pitching seasons of all time for the Detroit Tigers. In 1968, the Tigers' ace became the first 30-game winner since 1934, and the first in the American League since 1931. That year he definitively won both the American League MVP and Cy Young Award and is among 11 pitchers to win both prizes in the same year. In the National League, Bob Gibson, who also had an outstanding season in 1968 (for the St. Louis Cardinals), similarly won both the MVP and Cy Young Award. For this reason, 1968 is often referred to as "the year of the pitcher."

As great as Gibson's 1968 season was – a season in which he posted a remarkable 1.12 ERA – we believe that Denny McLain's season was even better. This is because McLain's 31 wins yielded a higher z-score (3.56) than any of Gibson's impressive stats that season, and it was also higher than Charley Radbourn's z-score of 2.57 in 1884, when he won 59 games. Hence, when it comes to comparing z-scores for victories in a single season, Denny McLain is the big *winner*.

Why Is Hugh Duffy's .440 Batting Average *NOT* Among the Top 20 Hitting Z-scores?

If there is a hitting counterpart to Charley Radbourn – whose remarkable 59 wins in 1884 didn't yield a z-score high enough to make our top-20 list for pitchers – it is probably Hugh Duffy. In 1894, Hugh Duffy of the Boston Braves posted a batting average of .440! By today's standards, this is hard to fathom, as it is even rare for a player to achieve an on-base average of .440. Yet, for all the splendor of this gargantuan batting average, his z-score for this incredible feat was just 2.90 and not close to making our top-20 list for hitters.

Just as high win totals were common in the Pre-Modern Era (and particularly in 1884, when Radbourn won 59 games and two other pitchers won at least 40 games), high batting averages were also common. Overall, ballparks were larger, which meant that outfielders were forced to cover more ground, thus making it easier for batters to find gaps in the field to get hits. Additionally, certain rule differences (such as four strikes being required for a strikeout) and scoring peculiarities (such as walks counting as hits in some seasons) made for a favorable environment for inflated batting averages. Hence, when Hugh Duffy hit .440 in 1894 he was among four players who hit over .400 that season and among 57 that hit over .300. In fact, the league mean for batting average in 1894 was a whopping .320! Retiring with a .324 lifetime average, Hugh Duffy is best known for recording the highest single-season batting average in baseball history, and he enjoyed 10 consecutive seasons of hitting at least .300, which helped him get elected to the Hall of Fame in 1945.

Though Hugh Duffy owns the record for the highest batting average in Major League history, the highest z-score for a batting average that we calculated was for Nap Lajoie. In 1901, Nap Lajoie of the Philadelphia A's posted an exceptional .426 average. However, while his raw batting average was just a few points lower than Hugh Duffy's mark of .440, Lajoie's z-score of 3.95 was significantly higher than Duffy's 2.90. It might not seem like there could have been such a big difference in baseball in the seven years between Duffy's .440 mark in 1894 and Lajoie's .426 mark in 1901, but rule changes and scoring changes made it more difficult to get hits in 1901. For example, the National League mean for batting average in 1901 was .294, as compared to .320 in 1894. So Lajoie put up a batting average that was almost as high as Duffy's historic record, but in an era when it much harder to get hits.

Lajoie, an infielder for 21 seasons, compiled more than 3,200 hits and was a three-time batting champion in the American League. He hit .300 or better 16 times over the course of his career and finished with a .328 lifetime average. He, like Charlie Radbourn, was elected to the Hall of Fame in 1939 in its inaugural year.

CHAPTER 4

PITCHERS WINNING THE MVP

"Pitchers are the dumbest guys on the field."

- Ted Williams,
Boston Red Sox

Comparing Pitching and Hitting Statistics Directly

The main benefit of using z-scores to analyze and evaluate performance, whether in baseball, academics or anything else, is that it enables you to compare things that are evaluated using different metrics or forms of assessment. As we've repeatedly stated throughout the book, z-scores allow you to compare *apples to oranges*, and in the context of baseball, this means that we can directly compare pitching and hitting statistics.

Why would this be useful? Well, we can think of numerous reasons, particularly for general managers and sports agents. They are responsible for determining the value of pitchers and hitters, sometimes in direct relation to each other for the purpose of contract negotiations and for making equitable trades. And then for the sportswriters who vote for individual postseason awards, and for the average fan (intent on proving that the sportswriters were wrong), there is another application that is even more relevant: determining whether a pitcher should win the Most Valuable Player (MVP) Award over a position player.

The MVP Award, which is currently issued to a single player in the American League and a single player in the National League, is officially known as the Baseball Writers' Association of America's Most Valuable Player Award, and it was first issued in 1931. Before 1931, other awards with different voting schemes were issued.[43] Since the inception of the MVP Award in 1931, 21 pitchers have won the award; however, between 1931 and 1956, the MVP Award was the only individual postseason award that a pitcher could win, as the Cy Young Award had not been established yet to honor the best pitchers in Major League Baseball.

Of the 21 pitchers to win the MVP, 11 won the award between 1931 and 1956, when it was the only major award issued, while 10 others have won the award since the inception of the Cy

---

[43] Gillette, Gary; Palmer, Pete (2007). *The ESPN Baseball Encyclopedia (Fourth ed.).*

Young Award in 1956. The focus of this chapter will be on these 10 pitchers who have won the MVP since the establishment of the Cy Young Award, as there has been a continuous debate within baseball circles during the past six decades about whether a pitcher should ever win the MVP when there is already an award to honor the best pitcher in each league.[44]

In throwing our hats into the ring of this lively debate, we decided to compare the z-scores of the 11 pitchers who have won the MVP since 1956, with the respective position player who achieved the next-highest vote total for the award in each respective season. In Table 4.1, we provide a list of these 11 pitchers who have won the MVP and their position player counterparts who had the next-highest vote total for the MVP in each respective season.

---

[44] Since 1967, the Cy Young Award has been issued to the best pitcher in each league; however, between 1956 and 1966, only one Cy Young Award was issued for the best pitcher in all of Major League Baseball.

TABLE 4.1: PITCHERS WHO HAVE WON THE MVP AWARD SINCE 1956 AND THEIR RESPECTIVE RUNNERS UP

YEAR	LEAGUE	MVP-WINNING PITCHER	MVP RUNNER-UP
1956	NL	Don Newcombe – Brooklyn Dodgers	Hank Aaron – Milwaukee Braves
1963	NL	Sandy Koufax – LA Dodgers	Dick Groat – St. Louis Cardinals
1968	NL	Bob Gibson – St. Louis Cardinals	Pete Rose – Cincinatti Reds
1968	AL	Denny McLain – Detroit Tigers	Bill Freehan – Detroit Tigers
1971	AL	Vida Blue – Oakland A's	Sal Bando – Oakland A's
1981	AL	Rollie Fingers – Milwaukee Brewers	Rickey Henderson – Oakland A's
1984	AL	Willie Hernandez – Detroit Tigers	Kent Hrbek – Minnesota Twins
1986	AL	Roger Clemens – Boston Red Sox	Don Mattingly – New York Yankees
1992	AL	Dennis Eckersly – Oakland A's	Kirby Pucket – Minnesota Twins
2011	AL	Justin Verlander – Detroit Tigers	Jacoby Ellsbury – Boston Red Sox
2014	NL	Clayton Kershaw – LA Dodgers	Giancarlo Stanton – Miami Marlins

We began our analyses by isolating the same statistical categories that we examined in our previous analyses (i.e., batting averages, home runs, RBI and OPS for position players, and wins, ERAs, strikeouts and WHIP for starting pitchers) and then created z-scores for those statistics[45] for the MVP pitchers and their respective position player runners-up. However, we made a few slight modifications in these analyses. First, since three of the 11 pitchers who won the MVP were relief pitchers (Rollie Fingers, Willie Hernandez and Dennis Eckersly), we decided to make a couple of substitutions regarding the statistical categories we examined. First, since relief pitchers are much more likely to acquire *saves* than they are to acquire *wins*, we substituted *saves* for *wins*. Second, since relief pitchers tend to pitch many fewer innings than starting pitchers, we decided to substitute *strikeouts per nine innings* for *total strikeouts*.

In the analyses for relief pitchers, we first had to determine criteria that we felt could accurately categorize a pitcher as a relief pitcher (as opposed to a starter). Since many starting pitchers occasionally have relief appearances during the regular season (and may also log a few saves), we decided that we could consider a pitcher to be a reliever if at least 51% of his appearances came in relief.[46] Additionally, since several of the position players who were runners-up for the MVP were not traditional sluggers, but utilized other weapons such as base stealing to boost their team's offensive output (such as Rickey Henderson and Jacoby Ellsbury), we

---

[45] As discussed in Chapter 2, we corrected the pitchers' Z-scores for ERA and WHIP, inverting them from negative numbers to positive numbers for ease of comparison.

[46] Most relief pitchers do not pitch enough innings to qualify for the ERA title (and were thus excluded from our previous analyses). In an effort to compute the most accurate analyses possible for the seasons in which a relief pitcher had won the MVP, we included in our data sets for those seasons all starting pitchers who qualified for the ERA title, as well as all relief pitchers who had at least as many appearances as the starting pitcher with the fewest total appearances.

calculated z-scores for stolen bases for those individuals in addition to the other hitting categories.

How to Determine Who Is the Best Player

Though we believe that the use of z-scores is a tremendous benefit in aiding the evaluation process of comparing hitters and pitchers by putting all of their relevant statistics on the same metric (i.e., the z-score metric), we were still left with a quandary in terms of how to determine which player is the best. Why? Because there are multiple z-scores that can be used to compare players. So we were left asking ourselves how we could determine who is the better player. Is it the player who has highest *single* z-score among all of the statistical categories we examined? Is it the player who has the highest *average* z-score across all of the categories we examined? Is the player who was most consistent and had the ***narrowest range*** of z-scores (from highest to lowest) across all of the categories we examined?

There are no right or wrong answers to these questions, which is why the determination of *who is best* is always fraught with subjective bias about what being "the best" actually means, and that's even before discussing the importance of intangible factors like clutch performance and leadership. Our analyses will not settle these debates, but the z-scores can provide baseball enthusiasts with the most accurate data for enriching the statistical side of the debate so that pitchers and hitters can be evaluated more equitably – on an even playing field, so to speak.

For the purposes of this book, we decided to present the data for all three evaluative criteria (i.e., the highest single z-score, the highest average z-score, and the narrowest range of z-scores) and make a qualitative judgment on a case-by-case basis. These data are presented below.

1956 NATIONAL LEAGUE MVP AWARD:
DON NEWCOMBE (MVP WINNER) VS. HANK AARON

Don Newcombe – Brooklyn Dodgers

	Wins	ERA	Strikeouts	WHIP
Raw Score	27	3.06	139	1.28
Z-score	2.73	1.10	0.93	2.47

Hank Aaron – Milwaukee Braves

	Batting Avg.	Home Runs	RBI	OPS
Raw Score	.328	26	92	0.923
Z-score	2.02	0.60	1.04	1.38

Highest Single Z-score: Don Newcombe (Wins) – 2.73
Highest Average Z-score Across Categories: Don Newcombe – 1.81
Narrowest Range of Z-scores: Hank Aaron – 1.42 (2.02 – 0.60)

Don Newcombe appears to be the clear winner in this comparison. Not only did Newcombe have the highest single z-score (2.73 for wins), but he had the highest average z-score across the categories we examined. Though Hank Aaron had a narrower range of z-scores, suggesting more consistent overall performance across skills, this seemed likely to be a function of his lower z-scores overall, including his having the *lowest* single z-score in our comparison – .60 for home runs – which is surprising given his reputation as a great home run hitter. We believe the BBWAA made the right decision in awarding the MVP to pitcher Don Newcombe over position player Hank Aaron.

1963 NATIONAL LEAGUE MVP AWARD:
SANDY KOUFAX (MVP WINNER) VS. DICK GROAT

Sandy Koufax – Los Angeles Dodgers

	Wins	ERA	Strikeouts	WHIP
Raw Score	25	1.88	306	0.87
Z-score	2.11	2.17	3.04	2.39

Dick Groat – St. Louis Cardinals

	Batting Avg.	Home Runs	RBI	OPS
Raw Score	.319	6	73	0.827
Z-score[47]	1.63	-0.79	0.32	0.88

Highest Single Z-score: Sandy Koufax (Strikeouts) – 3.04
Highest Average Z-score Across Categories: Sandy Koufax – 2.43
Narrowest Range of Z-scores: Sandy Koufax – 0.93 (3.04 – 2.11)

This comparison was not even close as Sandy Koufax swept all three of our evaluation categories. Koufax had the highest single z-score (3.04 for strikeouts) and the highest average z-score across the categories we examined. He also had a narrower range of z-scores, which is very impressive because all of his z-scores were higher than the highest of Dick Groat's z-scores. We believe the BBWAA made the right decision in awarding the MVP to pitcher Sandy Koufax over position player Dick Groat.

---

[47] Dick Groat's home run total (six) in 1963 was below the mean (14.22) for that season, and hence he has a negative Z-score, which we left uncorrected.

1968 NATIONAL LEAGUE MVP AWARD:
BOB GIBSON (MVP WINNER) VS. PETE ROSE

Bob Gibson – St. Louis Cardinals

	Wins	ERA	Strikeouts	WHIP
Raw Score	22	1.12	268	0.85
Z-score	2.11	2.76	2.61	2.46

Pete Rose – Cincinnati Reds

	Batting Avg.	Home Runs	RBI	OPS
Raw Score	.335	10	49	0.861
Z-score[48]	2.38	-0.13	-0.44	1.47

Highest Single Z-score: Bob Gibson – 2.76
Highest Average Z-score Across Categories: Bob Gibson – 2.46
Narrowest Range of Z-scores: Bob Gibson – 0.65 (2.76 – 2.11)

Bob Gibson dominated this comparison, sweeping all three of our evaluation categories. Gibson had the highest single z-score (2.76 for ERA) and the highest average z-score across the categories we examined. He also had a narrower range of z-scores, which is very impressive because all but one of his z-scores were higher than the highest of *Charlie Hustle's* z-scores. We believe the BBWAA made the right decision in awarding the MVP to pitcher Bob Gibson over position player Pete Rose.

---

[48] Pete Rose's home run total (10) and RBI total (49) in 1968 were both below the mean (11.36 and 58.73, respectively) for that season, and hence he had negative z-scores for those categories.

1968 AMERICAN LEAGUE MVP AWARD:
DENNY MCLAIN (MVP WINNER) VS. BILL FREEHAN

Denny McLain – Detroit Tigers

	Wins	ERA	Strikeouts	WHIP
Raw Score	31	1.96	280	0.90
Z-score	3.56	1.43	2.43	1.99

Bill Freehan – Detroit Tigers

	Batting Avg.	Home Runs	RBI	OPS
Raw Score	.263	25	84	0.819
Z-score	0.41	1.16	1.36	1.12

Highest Single Z-score: Denny McLain (Wins) – 3.56
Highest Average Z-score Across Categories: Denny McLain – 2.35
Narrowest Range of Z-scores: Bill Freehan – 0.95 (1.36 – 0.41)

Denny McLain was the clear winner of this comparison with the highest single z-score (3.56 for wins) and the highest average z-score across the categories we examined. Though McLain's teammate, Bill Freehan, had a narrower range of z-scores, suggesting more consistent overall performance across skills, this seemed likely to be a function of his lower z-scores overall, including his having the *lowest* single z-score in our comparison (0.41 for batting average). Hence, we believe the BBWAA made the right decision in awarding the MVP to pitcher Denny McLain over position player Bill Freehan.

1971 AMERICAN LEAGUE MVP AWARD:
VIDA BLUE (MVP WINNER) VS. SAL BANDO

Vida Blue – Oakland Athletics

	Wins	ERA	Strikeouts	WHIP
Raw Score	24	1.82	301	0.95
Z-score	1.97	2.38	2.81	2.15

Sal Bando – Oakland Athletics

	Batting Avg.	Home Runs	RBI	OPS
Raw Score	.271	24	94	.828
Z-score	0.09	0.93	1.39	0.80

Highest Single Z-score: Vida Blue (Strikeouts) – 2.81
Highest Average Z-score Across Categories: Vida Blue – 2.33
Narrowest Range of Z-scores: Vida Blue – 0.84 (2.81 – 1.97)

Vida Blue swept all three of our evaluation categories, as he had the highest single z-score (2.81 for strikeouts) and the highest average z-score across the categories we examined. He also had a narrower range of z-scores, and all of his z-scores were higher than all of teammate Sal Bando's. We believe the BBWAA made the right decision in awarding the MVP to pitcher Vida Blue over position player Sal Bando.

1981 AMERICAN LEAGUE MVP AWARD[49]:
ROLLIE FINGERS (MVP WINNER) VS. RICKEY HENDERSON

Rollie Fingers – Milwaukee Brewers

	Saves	ERA	Strikeouts / IP	WHIP
Raw Score	28	1.04	0.78	0.87
Z-score	3.85	2.91	1.57	2.48

Rickey Henderson – Oakland Athletics

	Batting Avg.	Home Runs	RBI	OPS	Stolen Bases
Raw Score	.319	6	35	.845	56
Z-score[50]	1.69	-0.40	-0.63	1.24	4.64

Highest Single Z-score: Rickey Henderson (Stolen Bases) – 4.64
Highest Average Z-score Across Categories: Rollie Fingers – 2.70
Narrowest Range of Z-scores: Rollie Fingers – 2.28 (3.85 – 1.57)

This comparison is the first to come up with mixed results and leave us unsure as to which player had the better season. The 1981 was one in which there was an abridged schedule as a result of the MLB players strike, which lasted from June 12 to August 9. As a result of the strike, the season was divided into two halves – the half-season before the strike and the half-season after the strike – and consequently, teams not only played fewer games (anywhere from 51 to 59 fewer), but not all teams played the same number of games. Some teams played as few as 103 and others played as many as 111. Because of this significant discrepancy, absolute statistics

---

[49] The 1981 season was shortened because of the MLB players' strike. The number of games played by each team ranged from 103 to 111.
[50] Rickey Henderson's home run total (six) and RBI total (35) in 1981 were both below the mean (8.55 and 44.27, respectively) for that season, and hence he had negative z-scores for those categories.

(e.g., total number of home runs) were not only much lower than they would have been during a full 162-game schedule, but performances on these statistics were much more difficult to assess.

Reliever Rollie Fingers won the MVP this season, and though his total number of raw saves – 28 – seems unimpressive at first glance (especially when compared to the single-season record of 62 saves by Francisco Rodriguez in 2008), his z-score for saves that season – 3.85 – is exceptionally high. Since his z-score was computed using the mean and standard deviation of the other relievers who pitched that season, all of whom were hampered by an abridged schedule because of the strike, his z-score for saves can be compared to those of any relief pitcher in any season.

Comparing his z-score for saves with Rickey Henderson's z-score for stolen bases is quite interesting and revealing. As we saw with Rollie Fingers' raw saves total in 1981, Rickey Henderson's raw stolen base total in 1981 does not seem exceptionally impressive (especially compared to the single-season record of 138 by Hugh Nicol in 1887, or Rickey Henderson's own total of 130 in 1982). But his z-score for stolen bases that season – 4.64 – is not only staggering, it is one of the top baseball z-scores of all-time and nearly a full z-score unit (i.e., a full standard deviation) above Rollie Fingers' impressive z-score for saves.

Compared to Henderson, what Fingers lacked in exceptional performance he made up for in consistency, as he had the higher average z-score across the categories we examined and also the more narrow range of z-scores. So this comparison has led to a split decision from a purely statistical perspective – one that could benefit from a more qualitative analysis for resolution, perhaps using an assessment of clutch performances. Nevertheless, given that Rollie Fingers led two of the three of *our* evaluation categories, we believe the BBWAA was on solid footing in awarding him the MVP over Rickey Henderson. We are sure Rickey Henderson would disagree… and would likely do so referring to himself in the third-person.

1984 AMERICAN LEAGUE MVP AWARD:
WILLIE HERNANDEZ (MVP WINNER) VS. KENT HRBEK

Willie Hernandez – Detroit Tigers

	Saves	ERA	Strikeouts / IP	WHIP
Raw Score	32	1.92	0.80	0.94
Z-score	2.29	2.30	1.21	2.59

Kent Hrbek – Minnesota Twins

	Batting Avg.	Home Runs	RBI	OPS
Raw score	.311	27	107	.906
Z-Score	1.26	0.98	1.39	1.55

Highest Single Z-score: Willie Hernandez (WHIP) – 2.59
Highest Average Z-score Across Categories: Willie Hernandez – 2.10
Narrowest Range of Z-scores: Kent Hrbek – 0.57 (1.55 – 0.98)

Willie Hernandez was the clear winner of this comparison with the highest single z-score (2.59 for WHIP) and the highest average z-score across the categories we examined. Though Kent Hrbek had a narrower range of z-scores, suggesting more consistent overall performance across skills, this seemed likely to be a function of his lower z-scores overall, including his having the *lowest* single z-score in our comparison (0.98 for home runs). We believe the BBWAA made the right decision in awarding the MVP to relief pitcher Willie Hernandez over position player Kent Hrbek.

1986 AMERICAN LEAGUE MVP AWARD:
ROGER CLEMENS (MVP WINNER) VS. DON MATTINGLY

Roger Clemens – Boston Red Sox

	Wins	ERA	Strikeouts	WHIP
Raw Score	24	2.48	238	0.97
Z-score	2.80	2.45	2.19	2.62

Don Mattingly – New York Yankees

	Batting Avg.	Home Runs	RBI	OPS
Raw Score	.352	31	113	.967
Z-score	2.74	1.33	1.74	2.45

Highest Single Z-score: Roger Clemens (Wins) – 2.80
Highest Average Z-score Across Categories: Roger Clemens – 2.52
Narrowest Range of Z-scores: Roger Clemens – 0.61 (2.80 – 2.19)

Though "The Rocket" was plagued by allegations of steroid use and faced charges for perjuring himself before Congress regarding his steroid use (charges from which he was eventually vindicated), he swept "Donnie Baseball" in this z-score comparison of their 1986 statistics. Hence, we believe the BBWAA made the right decision in awarding the MVP to Roger Clemens over Don Mattingly. It's a good thing George Steinbrenner isn't around to read this.

1992 AMERICAN LEAGUE MVP AWARD:
DENNIS ECKERSLY (MVP WINNER) VS. KIRBY PUCKET

Dennis Eckersly – Oakland Athletics

	Saves	ERA	Strikeouts / IP	WHIP
Raw Score	51	1.91	1.16	0.91
Z-score	3.81	1.50	2.96	2.13

Kirby Pucket – Minnesota Twins

	Batting Avg.	Home Runs	RBI	OPS
Raw Score	.329	19	110	.864
Z-score	2.57	0.40	1.64	1.28

Highest Single Z-score: Dennis Eckersly (Saves) – 3.81
Highest Average Z-score Across Categories: Dennis Eckersly – 2.60
Narrowest Range of Z-scores: Kirby Pucket – 2.17 (2.57 – 0.40)

Dennis Eckersly was the clear winner of this comparison, earning the highest single z-score (3.81 for saves) and the highest average z-score across the categories we examined. Although Kirby Pucket had a narrower range of z-scores, suggesting more consistent overall performance across skills, this seemed likely to be a function of his lower z-scores overall, including his having the *lowest* single z-score in our comparison (0.40 for home runs). We believe the BBWAA made the right decision in awarding the MVP to pitcher Dennis Eckersly over position player Kirby Pucket.

2011 AMERICAN LEAGUE MVP AWARD:
JUSTIN VERLANDER (MVP WINNER) VS. JACOBY ELLSBURY

Justin Verlander – Detroit Tigers

	Wins	ERA	Strikeouts	WHIP
Raw Score	24	2.40	250	0.92
Z-score	3.24	1.81	2.18	2.28

Jacoby Ellsbury – Boston Red Sox

	Batting Avg.	Home Runs	RBI	OPS	Stolen Bases
Raw Score	.321	32	105	.928	39
Z-score	1.70	1.34	1.37	1.56	2.20

Highest Single Z-score: Justin Verlander (Wins) – 3.24
Highest Average Z-score Across Categories: Justin Verlander – 2.38
Narrowest Range of Z-scores: Jacoby Ellsbury – 0.86 (2.20 – 1.37)

    Justin Verlander was the clear winner of this comparison with the highest single z-score (3.24 for wins) and the highest average z-score across the categories we examined. Though Jacoby Ellsbury had a narrower range of z-scores, suggesting more consistent overall performance across skills, this seemed likely to be a function of his lower z-scores overall, including his having the *lowest* single z-score in our comparison (1.34 for home runs). We believe the BBWAA made the right decision in awarding the MVP to pitcher Justin Verlander over Jacoby Ellsbury – especially when you consider that Verlander threw a no-hitter in 2011, as well.

2014 National League MVP Award:
Clayton Kershaw (MVP Winner) vs. Giancarlo Stanton

Clayton Kershaw – Los Angeles Dodgers

	Wins	ERA	Strikeouts	WHIP
Raw Score	21	1.77	239	0.86
Z-score	2.32	2.30	2.08	2.80

Giancarlo Stanton – Miami Marlins

	Batting Avg.	Home Runs	RBI	OPS
Raw Score	.288	37	105	.950
Z-score	0.73	2.63	1.95	2.35

Highest Single Z-score: Clayton Kershaw (WHIP) – 2.80
Highest Average Z-score Across Categories: Clayton Kershaw – 2.38
Narrowest Range of Z-scores: Clayton Kershaw – 0.72 (2.80 – 2.08)

Clayton Kershaw was the clear winner of this comparison with the highest single z-score (2.80 for WHIP), the highest average z-score, and the narrowest range of z-scores across the categories we examined. We believe the BBWAA made the right decision in awarding the MVP to pitcher Clayton Kershaw over Giancarlo Stanton – especially when you consider that Kershaw threw a no-hitter in 2014, as well.

Pedro Was Robbed! Twice!

According to our z-score analyses, we believe that in each of the seasons in which a pitcher has won the MVP, he was more deserving than each respective position player that placed second in the voting that season. Hence, it would *seem* that the BBWAA can handle the job of knowing when to choose a pitcher for the award without using our nifty little z-score. But, let's consider the other side of the MVP coin. Did the BBWAA get it right in each of the instances in which a position player won the MVP Award over a pitcher (which has occurred 85% of the time the award has been given)?

Given the z-scores reported in Chapter 3, the accomplishments of one pitcher – Pedro Martinez – during his two best seasons (1999 and 2000) piqued our interest for investigation. As noted in Chapter 3, Pedro Martinez achieved pitching feats in 1999 and 2000 for the Red Sox that we believe are more exceptional than those of any other pitcher in baseball history. Pedro's *raw* statistics in those seasons would be considered exceptional in any era, but the *z-scores* for those accomplishments are other-worldly.

Remember that Pedro's feats during the 1999 and 2000 seasons were achieved during the heart of the Steroid Era – the most offensively dominant era in baseball history. As such, we demonstrate with the z-score comparisons on the following page that, from a purely statistical perspective, the BBWAA robbed Pedro Martinez, not once, but twice for the MVP Award in 1999 and 2000, when it gave the award to position players Ivan Rodriguez and Jason Giambi, respectively.

Pedro Martinez was the clear winner of both of these comparisons with the highest single z-score in each season (**4.25** for strikeouts in 1999 and **4.47** for ERA in 2000) and the highest average z-score across the categories we examined in each respective season.

1999 AMERICAN LEAGUE MVP AWARD:
IVAN RODRIGUEZ (MVP WINNER) VS. PEDRO MARTINEZ

Pedro Martinez – Boston Red Sox

	Wins	ERA	Strikeouts	WHIP
Raw Score	23	2.07	313	0.92
Z-score	2.71	3.18	4.25	3.24

Ivan Rodriguez – Texas Rangers

	Batting Avg.	Home Runs	RBI	OPS
Raw Score	.332	35	113	.914
Z-score	1.58	1.15	0.97	0.73

Highest Single Z-score: Pedro Martinez (Strikeouts) – 4.25
Highest Average Z-score Across Categories: Pedro Martinez – 3.35
Narrowest Range of Z-scores: Ivan Rodriguez – 0.85 (1.58 – 0.73)

2000 AMERICAN LEAGUE MVP AWARD:
JASON GIAMBI (MVP WINNER) VS. PEDRO MARTINEZ

Pedro Martinez – Boston Red Sox

	Wins	ERA	Strikeouts	WHIP
Raw Score	18	1.74	284	0.74
Z-score	1.44	4.47	3.38	4.20

Jason Giambi – Oakland Athletics

	Batting Avg.	Home Runs	RBI	OPS
Raw Score	.333	43	137	1.123
Z-score	1.55	1.93	1.86	2.55

Highest Single Z-score: Pedro Martinez (ERA) – 4.47
Highest Average Z-score Across Categories: Pedro Martinez – 3.37
Narrowest Range of Z-scores: Jason Giambi – 1.00 (2.55 – 1.55)

Though Ivan Rodriguez and Jason Giambi had narrower ranges of z-scores, suggesting more consistent overall performance, in each case this seemed due to them having lower overall z-scores. In 1999, Ivan Rodriguez had the ***lowest*** single z-score (0.73 for OPS) in this head-to-head comparison with Pedro, and all of his z-scores were lower than all of Pedro's z-scores. In 2000, three-out-of-four of Pedro's z-scores were higher than the highest one of Jason Giambi's z-scores.

Given these two respective analyses, as well as the fact that in each of these two seasons Pedro Martinez achieved a z-score that ranks within the top eight overall z-scores we calculated (his 4.47 z-score for ERA in 2000 ranked 6th on our overall list and his 4.25 z-score for strikeouts in 1999 ranked 8th on our overall list), we believe the BBWAA wrongly gave the MVP Award to position players Ivan Rodriguez and Jason Giambi over Pedro Martinez in 1999 and 2000.

The analyses of this small sample suggest that there may have been many other cases in which the BBWAA overlooked the statistical feats of deserving pitchers, relative to position players, because there was previously no empirical way to compare pitchers' *apples* to position players' *oranges*. However, moving forward, we believe that if the BBWAA was to adopt the use of z-scores to compare statistics between position players and pitchers for the MVP Award, oversights like this (with Pedro Martinez getting snubbed, twice) would never happen again. Of course, we did not examine every possible statistic that may be considered in MVP voting, and the MVP Award is often about more than just quantitative statistical analysis, as intangibles, clutch performances, and team performance often play an important role in a player's selection for the award. Furthermore, our analyses do little to resolve the philosophical debate about whether a pitcher should ever win the MVP since there is already an award – the Cy Young Award – that honors the best pitcher in each league, each season. Nevertheless, our analyses suggest that from a purely statistical perspective, the BBWAA might be well-served in using z-scores to compare statistics between pitchers and position players for the MVP Award, and in a subsequent chapter we discuss how the BBWAA can use z-scores to better evaluate the worthiness of players in the Steroid Era for the Hall of Fame.

CHAPTER 5

REVENGE OF THE NERDS: AN HOMAGE TO *MONEYBALL*

*"It's about the money.
And when they say
it's not about the money,
then it is definitely about the money."*

- George Young, General Manager
NY Football Giants

Moneyball, the 2011 Academy Award-nominated film, based on Michael Lewis' 2003 bestseller, opens with a recap of the 2001 ALDS featuring the small-market Oakland A's (payroll $34 Million) against the wealthy, World Champion New York Yankees (payroll $110 Million): a classic David vs. Goliath matchup. Unfortunately for the Oakland A's and their high-strung general manager, Billy Beane (the protagonist of both the book and the movie), *this* David vs. Goliath contest didn't end like the one in the Bible (i.e., 1 Samuel, Chapter 17, for those of you keeping score at home), as Goliath beat David three games to two.

Moneyball details the machinations and unorthodox personnel decisions of A's general manager Billy Beane – a former five-tool phenom who played mostly in the Mets' minor league system – as he tries to find a way for his small-market team to compete with the wealthy titans of the American League (i.e., the Yankees, Red Sox, Angels and White Sox) despite having a fraction of the money to buy players. Interestingly, the Oakland A's were not always a frugal franchise. When the club was owned by Walter A. Haas, Jr. from 1980 through 1995, they were frequently among the American League's top teams in terms of payroll and had the highest payroll in the league in 1991. But with the death of Walter A. Haas, Jr. in 1995 and the eventual sale of the team, first to Stephen Schott and Ken Hofmann, and then to Lewis Wolff and Jordan Schaub, the Oakland A's have spent the succeeding two decades shopping in baseball's thrift stores with the rest of the league's lowest-payroll teams.

Despite general managing one of the league's lowest-payroll teams, Billy Beane has helped the A's achieve surprising success. During his tenure as general manager (from the 1998 season, through the 2014 season), the A's have had eleven winning seasons (twice winning over 100 games), making the playoffs eight times and winning the American League West title six times. Unfortunately, however, the A's have only won one playoff series since 1998 and have never made the World Series on Beane's watch.

How has Billy Beane managed to achieve such feats with the A's despite payroll restrictions forcing him to choose only from baseball's lowest-salaried players? This is the question that Michael Lewis explores in *Moneyball*. The answer is at once simple, and yet immeasurably complicated: he uses intricate statistics and mathematical models to target undervalued players with baseball skills that are most highly correlated with **team** success. As documented in *Moneyball*, Beane developed an appreciation for the esoteric statistics of baseball researcher, Bill James, from his predecessor and mentor, Sandy Alderson. James' research, which he began self-publishing annually in 1977 under the title of *The Bill James Baseball Abstract*, sought to determine which aspects of baseball performance were most highly correlated with team success: a.k.a. winning. What James discovered through his research was that many of the traditional baseball statistics that scouts, executives, historians and fans previously thought were instrumental to team success – such as batting average and RBI for hitters, and ERA and wins for pitchers – were not nearly as important as once believed. In their place, James averred that other factors – those most predictive of team run production and team run prevention – were more important because the object of baseball, quite simply, is for your team to score more runs in a given game than your opponent.

In place of some of the familiar statistics traditionally used to evaluate players, Bill James and other members of the Society for American Baseball Research (SABR) have touted hundreds of nouveau statistics that they believe to be more highly predictive of team success. Some of the most common include:

- **OPS** (On-Base-Average Plus Slugging Percentage)
- **wOBA** (Weighted On-Base-Average): wOBA is similar to the standard on-base average (i.e., times on base / plate appearances, multiplied by 1000) except each type of offensive event (i.e., hits, walks, hit-by-pitch, reaching base on an error, etc.) is weighted differently.
- **Runs Produced**: Runs Produced is the sum of Runs Scored, plus RBI, minus Home Runs.
- **WHIP** (Walks, plus Hits, divided by Innings Pitched)
- **WAR** (Wins Above Replacement): WAR is a complicated statistic estimating the number of wins a team has amassed as a result of a given player's total contribution (i.e., batting, base running, fielding and pitching), relative to a "replacement player" (i.e., a player of common skills, available for minimum cost).

The above list only covers a small fraction of sabermetric statistics devised and used by Bill James and others to help optimize team performance. For a more exhaustive list and comprehensive description – both, of the older sabermetric statistics, as well as the new statistics that are being developed every year – we encourage you to visit SABR's website at: http://sabr.org.

Making Chicken Salad Out of Chicken $#!+

The purpose of this book is to provide a context for the achievements and excellence of sport, particularly in baseball. Z-scores offer a way to not only assess achievement and excellence overall, but within the context in which a particular performance occurred. This is true not only for home runs and wins totals, but also for achievements like those of general managers, like Billy Beane, mentioned above. Without the appropriate context, it is impossible to know how to evaluate an achievement like Billy Beane general managing a team that made the playoffs seven out of 16 seasons with a payroll in the bottom half of the American League. In sports like hockey and basketball, where 16 out of 30 teams (i.e., 53%) in the league make the playoffs, this might not sound like much of an accomplishment. However, in baseball, where only eight of 30 teams (i.e., 27%) make the playoffs, this is much more significant. As such, we thought it would be interesting to provide a z-score context for several of Billy Beane's accomplishments as general manager of the Oakland A's.

Regarding the accomplishments of Billy Beane noted above, we will start by presenting z-scores for the number of winning seasons his team has had. In order to provide the appropriate context, we need to identify an appropriate reference group. For the first two analyses, the reference group we've identified as being most relevant is that of teams with payrolls for a given season that were below the league average for that season (i.e., these teams had a negative z-score for payroll in each season we examined). We thought this was an appropriate reference group because Billy Beane's genius has been to find ways to construct winning teams, year-after-year, with significantly less money to spend on players than most other teams. In Table 5.1, we identify the franchises that have had the most winning seasons during Billy Beane's tenure as general manager (from 1998 through 2013) despite having payrolls that were in the bottom half of their respective leagues.

TABLE 5.1: MOST WINNING SEASONS AMONG LOW PAYROLL TEAMS[51] FROM 1998-2014

RANK	TEAM	NUMBER OF WINNING SEASONS	LEAGUE MEAN	LEAGUE SD	Z-SCORE
1	Oakland A's	12	3.50	2.75	3.09
2	Minnesota Twins[52]	8	3.50	2.75	1.64
3	Toronto Blue Jays	8	3.50	2.75	1.64
4	Chicago White Sox	6	3.50	2.75	0.91
5	Tampa Bay Rays	6	3.50	2.75	0.91
6	Arizona Diamondbacks	5	3.50	2.75	0.55
7	Miami Marlins	5	3.50	2.75	0.55
8	San Diego Padres	5	3.50	2.75	0.55
9	(Five Teams Tied)	4	3.50	2.75	0.18

---

[51] Teams with a payroll that was less than the league mean for each given season examined.

[52] In 2010 the Twins had a winning record, posting 94 wins, however, their team payroll was above the league average that season, and hence, the Twins' eight winning seasons, as posted in this table, does not include the 2010 season.

TABLE 5.2: MOST PLAYOFF APPEARANCES AMONG LOW PAYROLL TEAMS[53] FROM 1998-2014

RANK	TEAM	NUMBER OF PLAYOFF APPEARANCES	LEAGUE MEAN	LEAGUE SD	Z-SCORE
1	Oakland A's	8	1.57	1.72	3.75
2	Minnesota Twins[54]	5	1.57	1.72	2.00
3	Tampa Bay Rays	4	1.57	1.72	1.42
4	(Nine Teams Tied)	2	1.57	1.72	0.25

As shown in Table 5.2, Billy Beane's Oakland A's have had the most winning seasons (12) since 1998 among teams with payrolls below the league mean, and this yielded an impressive z-score of 3.09. Though a z-score of 3.09 might not sound outstanding relative to some of the extraordinary z-scores reported in previous chapters, it is important to reiterate that a z-score of 2.0 is so impressive that when an individual has an IQ with a z-score of 2.0, that individual is eligible to be in Mensa, the society for people with high IQs. Even more impressive than the A's achievement for winning seasons under Billy Beane is their seven playoff appearances since 1998, yielding an even higher z-score of 3.75. Just to put this into context, a z-score of 3.75 would equate to **78 home runs** during the 2001 baseball season (when Barry Bonds hit 73 home runs). Interestingly, while Billy Beane and the Oakland A's have gotten all the acclaim over the past two decades for doing the most with the least, and rightfully so, right behind the Oakland

---

[53] Teams with a payroll that was less than the league mean for each given season examined.

[54] In 2010 the Twins made the playoffs, posting 94 wins, however, their team payroll was above the league average that season, and hence, the Twins' five playoff seasons, as posted in this table, does not include the 2010 season.

A's among low payroll teams, both in winning seasons and playoff appearances, is the Minnesota Twins. The Twins' general manager since 1994 has been Terry Ryan, and though Ryan has had nearly as much success as Billy Beane, his name isn't nearly as well-known outside of Minnesota or hardcore baseball circles.

Discussions about payrolls among baseball teams beg the question: Why do low payroll teams have low payrolls? Well, the answer has to do with many factors, including revenue from attendance and TV contracts, as well as the degree to which owners want to reinvest the revenue they collect to buy better talent or merely bank larger profits. Despite the myriad factors that influence actual team payrolls, two factors – market size (i.e., population of the city/region that supports a given team in terms of attendance and TV revenue) and market affluence seem the most highly correlated with team payrolls. While market size (i.e., population) is easy to assess, there are many possible ways to assess market affluence. The method that we present in this chapter is each baseball city's median household income. This was done for two reasons: first, these figures are easy to find in the US Census reports; and second, we believe this factor corresponds well with each team's maximum revenue potential, as it relates to demand for tickets (which correlates with ticket prices, and thus attendance revenue), as well as demand for TV broadcasting rights (which correlates with TV revenue potential in all its myriad forms).

Traditionally, assessments of the financial inequalities of the baseball landscape, distinguishing high payroll teams from low payroll teams, have focused solely on market size (i.e., population). However, we believe that this factor only explains part of the reason for the discrepancy between high and low payroll teams. To illustrate this point, the Red Sox, who traditionally have a team payroll within the top five of Major League Baseball, play in a city – Boston – with a population that is ranked 12th among Major League Baseball cities. However, in terms of median household income, Boston ranks 6th among Major League Baseball cities. Conversely, the Arizona Diamondbacks, who have traditionally been one of the lowest payroll teams in baseball, play in a city – Phoenix – that has a population in the top seven of Major League Baseball cities, but is ranked 12th in terms of median household income. In each case, with the Red Sox and the Diamondbacks, their team payrolls seem to correspond more with the median household income of the cities in which they play than the population of those cities. However, in many other cases, as with the Philadelphia Phillies, payroll size seems more highly correlated with population than with median household income. Our point is that both factors are important and should be jointly considered when distinguishing between "rich" teams and "poor" teams.

To account for both market size (i.e., population) AND market affluence (i.e., median household income), we created a variable called "market power" to reflect the joint contribution of both these factors. **Market power**, which we used to estimate the maximum money potentially available in a given market to support a baseball team, is simply the product of a market's population and its median household income. After calculating the market power product for each of Major League Baseball's 27 cities, we calculated z-scores for this new variable and these data are presented in Table 5.3. The derived data were obtained from the most recent US Census records for population and median household income.

TABLE 5.3: MARKET POWER OF MAJOR LEAGUE BASEBALL CITIES (BASED ON POPULATION AND MEDIAN HOUSEHOLD INCOME)[55]

RANK	CITY	2010 CITY POP.	2012 MEDIAN HOUSEHOLD INCOME	MARKET POWER[56] IN BILLIONS	MEAN IN BILLIONS	SD IN BILLIONS	Z-SCORE
1	New York, NY	8,175,133	$51,865	$424.00	$60.90	$87.50	4.15
2	Los Angeles, CA	3,792,621	$49,745	$188.66	$60.90	$87.50	1.46
3	Toronto, ON	2,615,060	$68,110	$178.11	$60.90	$87.50	1.34
4	Chicago, IL	2,695,598	$47,408	$127.79	$60.90	$87.50	0.76
5	Houston, TX	2,099,451	$44,648	$93.74	$60.90	$87.50	0.38
6	San Diego, CA	1,307,402	$63,990	$83.66	$60.90	$87.50	0.26
7	Phoenix, AZ	1,445,632	$47,866	$69.20	$60.90	$87.50	0.09
8	San Francisco, CA	805,235	$73,802	$59.43	$60.90	$87.50	-0.02
9	Philadelphia, PA	1,526,006	$37,016	$56.49	$60.90	$87.50	-0.05
10	Washington, DC	601,723	$64,267	$38.67	$60.90	$87.50	-0.25
11	Boston, MA	617,594	$53,136	$32.82	$60.90	$87.50	-0.32
12	Denver, CO	600,158	$49,091	$29.46	$60.90	$87.50	-0.36
13	Seattle, WA	608,660	$46,146	$28.09	$60.90	$87.50	-0.37
14	Baltimore, MD	620,961	$40,803	$25.34	$60.90	$87.50	-0.41
15	Milwaukee, WI	594,833	$35,823	$21.31	$60.90	$87.50	-0.45

---

[55] US Census: http://factfinder2.census.gov; accessed, October 22, 2014.
[56] "Market Power" refers to the product of a city's Population, multiplied by the city's Median Household Income. This figure represents an estimate of the city's potential purchasing power, and thus its ability to contribute to its sports teams' revenue through ticket sales, television rights revenue and other means.

TABLE 5.3 (CONTINUED)

RANK	CITY	2010 CITY POP.	2012 MEDIAN HOUSE-HOLD INCOME	MARKET POWER[57] *IN BILLIONS*	MEAN *IN BILLIONS*	SD *IN BILLIONS*	Z-SCORE
16	Kansas City, MO	459,787	$45,150	$20.76	$60.90	$87.50	-0.46
17	Oakland, CA	390,724	$51,683	$20.19	$60.90	$87.50	-0.47
18	Arlington, TX	365,438	$53,341	$19.49	$60.90	$87.50	-0.47
19	Atlanta, GA	420,003	$46,146	$19.38	$60.90	$87.50	-0.47
20	Detroit, MI	713,777	$26,955	$19.24	$60.90	$87.50	-0.48
21	Minneapolis, MN	382,578	$48,881	$18.70	$60.90	$87.50	-0.48
22	Tampa Bay, FL	335,709	$43,514	$14.61	$60.90	$87.50	-0.53
23	Miami, FL	399,457	$29,762	$11.89	$60.90	$87.50	-0.56
24	Pittsburgh, PA	305,704	$38,029	$11.63	$60.90	$87.50	-0.56
25	St. Louis, MO	319,294	$34,384	$10.98	$60.90	$87.50	-0.57
26	Cleveland, OH	396,815	$26,556	$10.54	$60.90	$87.50	-0.58
27	Cincinnati, OH	296,943	$33,708	$10.01	$60.90	$87.50	-0.58

---

[57] "Market Power" refers to the product of a city's Population, multiplied by the city's Median Household Income. This figure represents an estimate of the city's potential purchasing power, and thus its ability to contribute to its sports teams' revenue through ticket sales, television rights revenue and other means.

As you can see from Table 5.3, with regard to market power, the Oakland market is ranked 17th out of the 27 markets in Major League Baseball, yielding a z-score of -0.47, which indicates that Oakland's market power is below the league mean for all baseball markets. Despite playing in such a weak market, only two teams – the New York Yankees and the Boston Red Sox – have had more winning seasons (16 and 15, respectively) and playoff appearances than the Oakland A's during Billy Beane's tenure as general manager.

Interestingly, our research identified two other teams in baseball – the Atlanta Braves and the St. Louis Cardinals – that play in weaker markets than Oakland, but have had more winning seasons and playoff appearances than the A's under Billy Beane's tenure. Between 1998 and 2014, the Cardinals had 14 winning seasons and 11 playoff appearances while the Braves had 14 winning seasons and 10 playoff appearances (as compared to the A's, with 12 winning seasons and eight playoff appearances). However, despite playing in weaker markets, the Braves and Cardinals have consistently had higher payrolls than the A's for virtually all of the seasons that Beane has been the A's general manager. Why is this so? The reason is unclear, though it may relate to differences between the teams' owners in their willingness to either reinvest revenue to improve personnel or to simply increase their profit margins. Regardless, Billy Beane has consistently been forced to do more with less, and thus, deserves immense credit for being the best culinary wizard in baseball: the chef who is best able to make chicken salad out of chicken $#!+.

Salaries Between Sports

To our knowledge, applying z-scores to player salaries across the four major American team sports – baseball, basketball, football and hockey – has not been presented in published form before the printing of this book. Nevertheless, we believe that such analyses, and the comparisons made therefrom, offer many illuminating insights about the economics of each respective sports league. Table 5.4 below lists the average salaries for players in the four major sports, as well as the salary for the highest paid player in each sport, and the z-score for the highest salary in each respective sport.

TABLE 5.4: Z-SCORES FOR SALARIES OF THE HIGHEST PAID PLAYERS IN THE FOUR MAJOR AMERICAN TEAM SPORTS FOR THE 2013-2014 SEASON[58]

LEAGUE	NUMBER OF PLAYERS IN THE LEAGUE	HIGHEST PAID PLAYER IN THE LEAGUE	SALARY FOR HIGHEST PAID PLAYER FOR THE 2013-2014 SEASON	MEAN SALARY FOR THE LEAGUE FOR THE 2013-2014 SEASON	SD	Z-SCORE
MAJOR LEAGUE BASEBALL (MLB)[59]	1076	ALEX RODRIGUEZ	$28,000,000	$3,026,309	$4,470,252	5.60
NATIONAL FOOTBALL LEAGUE (NFL)[60]	2349	AARON RODGERS	$22,000,000	$1,780,041	$2,605,934	7.76
NATIONAL BASKETBALL ASSOCIATION (NBA)	444	KOBE BRYANT	$24,250,000	$4,376,897	$4,829,375	4.12
NATIONAL HOCKEY LEAGUE (NHL)	769	ALEX OVECHKIN	$9,538,462	$2,529,747	$1,971,301	3.56

As shown in Table 5.4, the highest paid athlete across the four major sports[61] during the 2013-2014 season was Alex Rodriguez, whose salary for the New York Yankees was

---

[58] www.sportrac.com
[59] Data refer only to the 2013 MLB season.
[60] Data refer only to the guaranteed portion of the player's contract.
[61] At the time of publication of this book.

$28,000,000 per year. However, despite being the highest paid athlete, Alex Rodriguez's z-score for his salary (5.60) isn't even close to Aaron Rodgers' z-score (7.76) for his more *modest* salary of $22,000,000 per year. Why? A closer look at the table shows that the average player salary in baseball is nearly twice as much as the average player salary in football. For this reason, the NFL, more than any other major sports league, seems to resemble a banana republic, starkly divided between the "haves" (i.e., the star players) and the "have nots," (i.e., the non-star players). In contrast to the NFL is the NHL, which has a much smaller discrepancy between the league's average player salary and that of its highest-paid players. During the 2013-2014 NHL season, the league's highest paid player, Alex Ovechkin, made less than half the salary of Aaron Rodgers, yet the average player salary in the NHL that season was nearly 1.5 times higher than the average player salary in the NFL. For this reason, it seems that the NHL is the most egalitarian of the four major sports leagues. Of course, based on the data in Table 5.4, if you are just an "average" player, the best league for you to play in is the NBA, which has the highest average player salary among the four major sports.

Going back to Table 5.4 and looking strictly at z-scores, a couple of interesting comparisons come to light. If a player making Alex Rodriguez's salary ($28 million) played in the NHL, his salary would have a z-score of 12.92; however, a baseball player making the same salary as Alex Ovechkin ($9.54 million), the highest paid NHL player, would only have a z-score of 1.46; and a basketball player with Alex Ovechkin's salary would only have a z-score of 1.07.

A player's single-season salary is just one of the factors related to his career earning potential: the other major factor is the number of years he plays in the league. As shown in Table 5.5, MLB players have the longest average career length (5.6 years), with players in the NHL and NBA close behind them. However, players in the NFL have careers that are considerably shorter than those in all of the other major sports (3.5 years). NFL players have such short careers, partly due to the intense physicality of this contact sport, which leads to many more acute and chronic injuries. As you can also see from Table 5.5, when it comes to average career earning potential (which we calculated by multiplying the average salary for each league with the average career length), NFL players are clearly the worst off, while players in the NBA are, by far, the best off. One of the main reasons for this is that there are far fewer players in the NBA (444) than there are in the NFL (2349), and thus, there are far fewer shares dividing the league's revenue pot.

If there is an overarching message from the data of this section it is probably that if you can choose a sport to play professionally, there are several reasons why football should not be your first choice. First, NFL players have significantly lower career earning potential. Second, most contracts for players in the NFL are non-guaranteed, which means that teams can cut them (and not pay them) for virtually any reason (including injury), any time they want. Furthermore, in recent years a growing number of former NFL players have been diagnosed with Chronic Traumatic Encephalopathy (CTE): a degenerative brain condition that leads to dementia, memory loss, depression and aggression, caused by frequent head trauma. Hence, if you combine the potential for CTE (and other physical injuries), which is higher in football than any of the other sports, with the relatively modest career earning potential in the NFL, it begs one to ask whether the costs of playing professional football outweigh the benefits.

TABLE 5.5: AVERAGE CAREER LENGTH AND CAREER EARNING POTENTIAL FOR PLAYERS IN THE FOUR MAJOR AMERICAN TEAM SPORTS

LEAGUE	AVERAGE CAREER LENGTH IN YEARS[62]	AVERAGE SALARY FOR THE LEAGUE FOR THE 2013-2014 SEASON	AVERAGE CAREER EARNING POTENTIAL
MAJOR LEAGUE BASEBALL (MLB)	5.6	$3,026,309	$16,947,330
NATIONAL FOOTBALL LEAGUE (NFL)	3.5	$1,780,041	$6,230,144
NATIONAL BASKETBALL ASSOCIATION (NBA)	4.8	$4,376,897	$21,009,106
NATIONAL HOCKEY LEAGUE (NHL)	5.5	$2,529,747	$13,913,609

---

[62] The USA Today: Sports - http://ftw.usatoday.com/2013/10/average-career-earnings-nfl-nba-mlb-nhl-mls/

The California Penal League

From the 1989 motion picture, *Major League*:

Willie Mays Hayes: What the hell league you been playing in?
Rick "Wild Thing" Vaughn: California Penal...
Willie Mays Hayes: Never heard of it. How'd you end up playing there?
Rick "Wild Thing" Vaughn: Stole a car.

Before Charlie Sheen was "winning," he was saving games as the character *Wild Thing* in the movie *Major League*. In the film, the Indians general manager is assigned the task of putting together a ragtag team of unknown players: several of whom were playing in leagues that were not quite *prime time*. Though the film makes a farce out of player scouting and selection, in truth, major league general managers and scouts are assigned the task of evaluating amateur and minor league players who play in leagues that are vastly different and often unfamiliar. This is especially true when evaluating high school players as there are significant differences, from league to league, in competition level and the size of playing fields. Differences in competition level and ballpark size also complicate evaluations of college and minor league players, but the problem of evaluating statistics is most pronounced at the high school level.

Z-scores, however, can be a tremendous aid for baseball general managers, particularly when evaluating amateur talent. As discussed in the previous sections regarding major league talent, raw statistics mean very little unless you have a context for those statistics, and z-scores provide that context. Based on geography, climate, level of competition and other idiosyncratic factors, it may be more or less easy to hit home runs or record strikeouts in one high school league or another. However, these factors are roughly similar for all players within each given league (i.e., for all data points in the distribution), allowing one to compute a unique mean and standard deviation for each league. These components – the league mean and standard deviation – provide the context for each raw statistic in that league, and are incorporated into each z-score.

Of course, general managers often have to evaluate players across myriad levels of play – especially when drafting amateur players or signing free agents. In today's world of baseball, general managers not only have to evaluate players from American high schools, colleges and minor league systems, but professional leagues in Latin America, Asia and even Europe: all with their own unique characteristics, season lengths and competition levels. Using z-scores, however, a general manager who may have to choose between drafting a Florida high school player with 10 home runs, a Stanford University player with 15 home runs and a South Korean pro with 30 home runs, can put all these statistics on an *even playing field* by converting them to z-scores. Z-scores are not the be-all end-all tool that will help general managers in *all* instances of player evaluations, but for general managers who take more of a sabermetric approach to player evaluation (as opposed to a more qualitative approach, based on observation and scouting), z-scores can be a very helpful way to make sense of disparate statistics among players in different leagues.

Moneyball's Moment of Zen

In conclusion, the sabermetric approach documented in *Moneyball* is both Zen and Anti-Zen. It is Zen in that it relies on *pragmatic* statistics, like on-base percentage, that are highly correlated with team success, rather than statistics like batting average, which, while less correlated with team success, have greater sentimental and traditional appeal among old-school scouts and fans.

On the flip side, the sabermetric approach is anti-Zen in that it relies too much on intellectualism and reason, and generally dismisses the use of gut instincts as a form of *ancient religion*,[63] rendering it irrelevant simply because it is not empirical. This is not just the case for those carrying the torch of sabermetrics, it is also true with all forms of empirical reductionism. We humans amount to much more than what the *observable* data about us suggest. In the context of baseball, things such as heart, grit, leadership, work ethic, winning instinct, emotional calm in pressure situations and dozens of other *Jeteresque* traits cannot be easily quantified, and thus, they cannot be easily studied empirically. But these traits are very important for building winning teams, and good scouts know in their gut how to assess these traits and predict the extent to which amateur players will use these traits to improve through their baseball careers.

The problem with the old school baseball guys that were lampooned in *MoneyBall* is not that they relied on their gut instincts, but that they relied TOO MUCH on their gut instincts, without realizing that they had become rusty and unwilling to adapt to changes in the game. The gut needs to be trained by experience, but the type of experiences that are relevant changes over time. To us, Billy Beane is a revolutionary, not merely because of his innovative use of sabermetrics, but because he had the chutzpah to tell the emperors (i.e., the scouts) that they weren't wearing any clothes; that the experiences leading to their gut instincts were becoming obsolete because the game was changing more quickly than they realized. However, despite Billy Beane's success, it is ironic, or perhaps not so to some, that his teams have neither won a World Series, nor a league championship in the 16 years of his tenure, despite other small market teams accomplishing this feat in the same time. This begs one to wonder whether his approach is overly reliant on empiricism to the exclusion of intangible, gut-oriented assessments.

---

[63] A reference to *Star Wars,* in which an Imperial officer mocks Darth Vader's belief in "the force", saying: "Your sad devotion to that ancient religion has not given you the ability to conjure up the stolen data plans!"

CHAPTER 6

Z-SCORES & OTHER SPORTS

"A good hockey player plays where the puck is. A great hockey player plays where the puck is going to be."

- Wayne Gretzky,
Hockey Hall of Famer

When it comes to sports statistics, for decades the two numbers that were etched in most peoples' minds were 61 and 755, which, of course respectively correspond to the single-season home run mark set by Roger Maris in 1961, and, the career home run record that Hank Aaron finished setting in 1976. Both of these records have since been eclipsed by Barry Bonds – who hit 73 home runs in 2001, and who finished his career in 2007 with 762 home runs – but other records still remain, carved into the stump of our collective sports consciousness: Joe DiMaggio's 56-game hitting streak; Orel Hershiser's streak of 59 consecutive scoreless inning pitched; Cal Ripken Jr.'s record of playing in 2,632 consecutive games; Cy Young's 511 career wins; Nolan Ryan's 5,714 career strikeouts and seven no-hitters; and Pete Rose's 4,256 hits.

Unlike any other sport, the statistical records in baseball seem much more memorable than those of football, basketball and hockey. Nevertheless, there have been many significant achievements and notable single-season records in each of those other sports. In this chapter we *truly* compare apples to oranges by examining how the z-scores for outstanding achievements in the other major sports compare to the accomplishments of baseball's stars highlighted in previous chapters. In contrast to our analyses for baseball, however, we did not divide the history of each sport into separate eras and stratify our data according to those eras. Rather, we simply identified the greatest single-season performances of all-time for a few statistical categories in each sport. The reason for taking this abridged approach is because each sport would require an entire book to delineate its history and stratified statistics by era. We thus decided to leave that work for future books of this series and instead whet the appetite of sports fans with a small sample of data from the other major sports.

Among the four major sports, baseball has the most rigorous criteria for qualifying league leaders in various categories, such as batting average. Currently, Major League Baseball requires that batters average 3.1 plate appearances per team game (for a total of

502 plate appearances for a given season, assuming a standard 162-game season) to qualify for the batting title. Less stringent criteria were used in previous eras. However, in an effort to keep comparisons on an *even playing field* between sports, we decided to use the least restrictive standard to create the list of qualifiers for each statistical category for each sport. As such, we determined that the least restrictive standard employed by MLB in its long history to determine a batting champion was participation in at least 65% of the team's games: this was the standard used between 1892 and 1949 in the National League. For the majority of seasons in that time span (i.e., 1892 – 1949), the schedule consisted of 154 games, and in order to qualify for the batting title, players needed to play in at least 100 games (i.e., 65% of their team's games).

We decided to apply this rule (65% of the team's games played) to the other sports to limit the pool of prospective players for each statistical category. The only exception that we made was for the two goaltending categories in hockey, since players have been credited with various single-season awards and statistical accolades having played in fewer than 50% of their team's games (e.g., Brian Elliot was credited with the highest save percentage in the 2011-2012 season, having played in only 46% of his team's games). For goaltending in hockey, we therefore decided to include all goaltenders in our analyses. As you peruse the tables on the following pages, remember that the baseball z-score to beat is **5.70**, which Babe Ruth earned for his 29 home runs in 1919.

Football Statistics[64]

Despite baseball owning a special place in our hearts as America's pastime, professional football has eclipsed baseball in terms of overall popularity, television ratings and generated revenue. Yet, for all of football's popularity, the statistical marks and records of the sport don't seem to roll off our tongues as easily as those of baseball. Nevertheless, there are some memorable records in professional football, and other records that are important markers of individual success for the sport. For this chapter, we calculated z-scores for the following single-season offensive and defensive statistics: total rushing yards, touchdown passes, passing completion percentages, total receiving yards, touchdown receptions, total sacks and total interceptions.

TABLE 6.1: Z-SCORES FOR THE NFL'S TOP SINGLE-SEASON PERFORMANCES

STATISTIC	PLAYER	YEAR	TEAM	RAW VALUE	LEAGUE MEAN (SD)	Z-SCORE
RUSHING YARDS	ERIC DICKERSON	1984-1985	LOS ANGELES RAMS	2105	402.30 (391.30)	4.35
TOUCHDOWN PASSES	PEYTON MANNING	2013-2014	DENVER BRONCOS	55	23.85 (9.44)	3.30
COMPLETION PERCENTAGE	DREW BREES	2011-2012	NEW ORLEANS SAINTS	71.2%	60.13% (4.98)	2.22
RECEIVING YARDS	CALVIN JOHNSON	2012-2013	DETROIT LIONS	1964	330.30 (350.70)	4.66
TOUCHDOWN RECEPTIONS	RANDY MOSS	2007-2008	NEW ENGLAND PATRIOTS	23	2.02 (2.93)	7.15
SACKS	MICHAEL STRAHAN	2001-2002	NY GIANTS	22.50	1.91 (2.84)	7.25
INTERCEPTIONS	NIGHT TRAIN LANE	1952-1953	LOS ANGELES RAMS	14	2.96 (2.43)	4.54

---

[64] All data used to compute these football Z-scores were obtained from http://www.pro-football-reference.com

Basketball Statistics[65]

For this comparison, we calculated z-scores for the following single-season basketball statistics: total points, points per game (PPG), total rebounds, rebounds per game (RPG), total assists, assists per game (APG) and three-point field goal percentage.

TABLE 6.2: Z-SCORES FOR THE NBA'S TOP SINGLE-SEASON PERFORMANCES

Statistic	Player	Year	Team	Raw Value	League Mean (SD)	Z-Score
Total Points	Wilt Chamberlain	1961-1962	Philadelphia Warriors	4029	962.21 (676.17)	4.54
Points Per Game (PPG)	Wilt Chamberlain	1961-1962	Philadelphia Warriors	50.36	12.83 (8.44)	4.45
Total Rebounds	Wilt Chamberlain	1961-1962	Philadelphia Warriors	2149	501.93 (372.79)	4.15
Rebounds Per Game (RPG)	Wilt Chamberlain	1961-1962	Philadelphia Warriors	27.20	6.71 (4.68)	4.05
Total Assists	John Stockton	1990-1991	Utah Jazz	1164	190.57 (176.06)	5.53
Assists Per Game (APG)	John Stockton	1990-1991	Utah Jazz	14.54	3.20 (2.62)	4.33
Three-Point Field Goal %	Kyle Korver	2009-2010	Utah Jazz	53.64%	28.25% (14.48)	1.74

---

[65] All data used to compute these basketball Z-scores were obtained from http://www.basketball-reference.com

Hockey Statistics[66]

For hockey, we calculated z-scores for the following single-season offensive and goaltending statistics: total goals, total assists, total points, total wins, total shutouts, goals against average (i.e., goals per game allowed) and save percentages.[67]

TABLE 6.3: Z-SCORES FOR THE NHL'S TOP SINGLE-SEASON PERFORMANCES

STATISTIC	PLAYER	YEAR	TEAM	RAW VALUE	LEAGUE MEAN (SD)	Z-SCORE
TOTAL GOALS	WAYNE GRETZKY	1981-1982	EDMONTON OILERS	92	18.58 (13.90)	5.28
TOTAL ASSISTS	WAYNE GRETZKY	1985-1986	EDMONTON OILERS	163	28.47 (17.23)	7.81
TOTAL POINTS	WAYNE GRETZKY	1985-1986	EDMONTON OILERS	215	45.87 (27.43)	6.17
GOALTENDING WINS	MARTIN BRODEUR	2006-2007	NEW JERSEY DEVILS	48	23.10 (11.71)	2.13
GOALTENDING SHUTOUTS	GEORGE HAINSWORTH	1928-1929	MONTREAL CANADIANS	22	9.08 (5.96)	2.17
GOALS AGAINST AVERAGE (GAA)	GEORGE HAINSWORTH	1928-1929	MONTREAL CANADIANS	0.92	1.49 (0.55)	1.04
SAVE %	BRIAN ELLIOT	2011-2012	ST. LOUIS BLUES	94.03%	91.40% (0.01)	2.38

---

[66] All data used to compute these hockey z-scores were obtained from http://www.hockey-reference.com

[67] As with ERA and WHIP (as discussed in Chapter 3), we corrected the z-score for Goals Against Average (GAA) for ease of comparison with the z-scores of other statistical categories by inverting the negative values into positive values. Normally, better performance in GAA would be associated with a more negative z-score, however, in our presentation, higher z-scores indicate better GAA performance.

TABLE 6.4: TOP 10 Z-SCORES FOR SINGLE-SEASON PERFORMANCES ACROSS THE FOUR MAJOR AMERICAN TEAM SPORTS

STATISTIC (SPORT)	PLAYER	YEAR	TEAM	RAW VALUE	LEAGUE MEAN	Z-SCORE
TOTAL ASSISTS (HOCKEY)	WAYNE GRETZKY	1985-1986	EDMONTON OILERS	163	28.47 (17.23)	7.81
SACKS (FOOTBALL)	MICHAEL STRAHAN	2001-2002	NEW YORK GIANTS	22.50	1.91 (2.84)	7.25
TOUCHDOWN RECEPTIONS (FOOTBALL)	RANDY MOSS	2007-2008	NEW ENGLAND PATRIOTS	23	2.02 (2.93)	7.15
TOTAL POINTS (HOCKEY)	WAYNE GRETZKY	1985-1986	EDMONTON OILERS	215	45.87 (27.43)	6.17
HOME RUNS (BASEBALL)	BABE RUTH	1919 (AL)	BOSTON RED SOX	29	3.47 (4.48)	5.70
TOTAL ASSISTS (BASKETBALL)	JOHN STOCKTON	1990-1991	UTAH JAZZ	1164	190.57 (176.06)	5.53
TOTAL GOALS (HOCKEY)	WAYNE GRETZKY	1981-1982	EDMONTON OILERS	92	18.58 (13.90)	5.28
STRIKEOUTS (BASEBALL)	DAZZY VANCE	1924 (NL)	BROOKLYN DODGERS	262	65.62 (38.27)	5.13
OPS (BASEBALL)	BARRY BONDS	2004 (NL)	SAN FRANCISCO GIANTS	1.422	0.824 (0.119)	5.02
HOME RUNS (BASEBALL)	GAVVY CRAVATH	1915 (NL)	PHILADELPHIA PHILLIES	24	3.82 (4.21)	4.79

Surpassing *The Babe*? Only in *Wayne's World*

Babe Ruth enjoys a mythical status in the United States and especially among baseball fans. He is both literally and figuratively the poster child for the sport of baseball and, some might argue, for American sports. In earlier chapters, we reported that Babe Ruth's z-score of 5.70 for home runs in 1919 was the highest z-score we computed for the sport of baseball. But it is not the highest z-score that we computed for this book. As you can see from Table 6.4, the z-score champion, and arguably the best athlete among all of North America's most popular team sports, is Wayne Gretzky. That's right: it's not the Babe.

Gretzky had not one, but two z-scores that were significantly higher than Babe Ruth's highest z-score of 5.70: a z-score of 7.81 for assists in 1985-86, and a z-score of 6.71 for points that same season. Moreover, along with his z-score of 5.28 for scoring 92 goals in 1981-82, Wayne Gretzky has three z-scores on our top-10 list covering the four major North American team sports.

To put into context just how outstanding Gretzky's z-score of 7.81 for assists is, this would equate to an individual with an IQ of 217 and a man who is nearly 7-foot-8-inches tall! Furthermore, if we bring the conversation back to home runs, particularly during Barry Bonds' record-setting season of 2001, Wayne Gretzky's z-score of 7.81 would translate to 138 home runs for that season!

Though hockey is clearly fourth in popularity among the four major team sports in the United States, its status in Canada (hockey's birthplace) is closer to that of a national religion. Wayne Gretzky, nicknamed "The Great One," is almost god-like in Canada. How great was Gretzky? During one point in his career, he owned or shared more than 50 National Hockey League records. But *The Great One* not only achieved tremendous individual accomplishments, he also led his Edmonton Oilers team to four Stanley Cup championships. In addition, Wayne Gretzky was one of hockey's greatest ambassadors, and most hockey experts agree that the two expansion teams the NHL added in the 1990s in southern California – the San Jose Sharks in 1991 and the Anaheim Mighty Ducks in 1993 – were a direct response to the popularity that hockey achieved in that region after Wayne Gretzky was traded to the Los Angeles Kings in 1988.

When Gretzky retired in 1999 from the New York Rangers, he was immediately inducted into the NHL Hall of Fame, bypassing the usual three-year waiting period; and at the 2000 NHL All-Star Game, Commissioner Gary Bettman announced that Gretzky's iconic number 99 would be retired by every NHL team, making him the second athlete in North American sports (behind Jackie Robinson) to have his uniform number retired league-wide. On a personal note, Wayne Gretzky was so popular in the Cottone household that my father named our dog "Gretzky" (inspired by the movie *Rocky,* where the eponymous character named his dog "Butkus" after the famous Bears linebacker). It was always a treat to watch *our* Gretzky bark at the TV whenever he would hear his name invoked with excitement during a hockey game.

As shown in Table 6.4, Gretzky was not the only athlete to achieve a z-score higher than Babe Ruth's 5.70 mark. Football player Michael Strahan achieved a z-score of 7.25 for sacks in 2001-2002; and football player Randy Moss achieved a z-score of 7.15 for touchdown receptions in 2007-2008. This just goes to show that statistics, while relevant and significant, are not the most important part of sports. Regardless of statistics – and even z-scores – it is inconceivable that any North American athlete will ever replace Babe Ruth in American culture as the Sultan of *sport*.

CHAPTER 7

HALL OF FAME ELIGIBILITY

"Honor means that a man is not exceptional; fame, that he is. Fame is something that must be won; honor, only something which not be lost."

- Arthur Schopenhauer,
German Philosopher

With the rise of steroid use in sports over the past 30 years, baseball fans, scouts, sportswriters and talk radio hosts have all been forced to account for the ways in which the use of these performance-enhancing drugs (PEDs) have artificially inflated individual players' statistics and also determine how players in the so-called Steroid Era should be evaluated regarding their worthiness for the Hall of Fame. Of course, steroid use is only the latest factor to change the game and make significant alterations to the statistics of individual players. As we noted in previous chapters, many other factors have changed the way baseball has been played across the eras of baseball history, leading to anomalies and aberrations in player statistics that have been at least as significant as those stemming from steroid use. To review, some of these include:

- Changes in the ball, from a "dead" ball to a "live" ball at the beginning of the 1920s… and eventually a "juiced" ball in the 1990s;
- The prohibition of players of color, followed by the admission of players of color after Jackie Robinson's inaugural season in 1947;
- Changes in the size of ballparks league-wide over the years from large, cavernous ballparks favoring pitchers and singles hitters in previous eras, to the smaller bandbox fields favoring power hitters today;
- Changes in the number of games played each season – from 70 in 1876, to 140 in 1900, to 154 in 1904, to 162 in 1962;
- Changes in pitching rotations – from one-to-two-man rotations in the Pre-Modern Era, to four-man rotations throughout much of the 20th century, and then to the five-man rotations, which have been the norm since the 1990s;
- The rise of pitching specialists in last half of the 20th century, including the specialized use of closers, set-up men, middle-relief pitchers and left-handed specialists (i.e., left-handed

relievers who are used almost exclusively against left-handed hitters);
- The advent of the DH in the American League in 1973;
- The expansion of the league from eight teams in 1876, to 30 teams today.

Each of these changes has led to significant variance in the range of statistics for individual players during the 140-year history of baseball, yet, there still seem to be fixed benchmarks for entry into the Hall of Fame that the Baseball Writers' Association of America (BBWAA) still abide by, such as: 500 home runs, 300 wins and 3,000 hits. The BBWAA has never accounted for the fact that the ease with which a player can hit a home run has been altered dramatically over the years by the lengthened schedule, watered-down pitching, rise of smaller ballparks and advent of the DH. Similarly, they have never accounted for the fact that it is now much more difficult for a starter to win 300 games, for two main reasons: First, with the change from four-man rotations to five-man rotations, starters get 20% fewer chances to get a win than they had in previous eras; and second, managers are more likely to pull their starters earlier in games (before they might get the chance to get a win) thanks to the specialized use of relief pitchers. Granted, these arbitrary benchmarks are not the only criteria used to evaluate the worthiness of a player for the Hall of Fame, but they are still granted significance despite their increasing irrelevance.

It is our hope that the data and arguments provided in this book can help change the debate for the Hall of Fame – especially as players from the so-called Steroid Era continue to retire and we are forced to find a context for their statistics. Z-scores can clarify how to handle these players, currently in purgatory, to help evaluate them against their peers from current and former eras, as well as existing Hall of Famers. We believe that the use of z-scores (particularly as we computed them in this book) in the evaluation process for the Hall of Fame is invaluable, as it automatically controls for all of the differences that exist between eras and seasons in baseball history. This is because each z-score uniquely reflects each player's performance in a given statistical category *relative to his peers* – peers who played in the same league and same season, and were thus subjected to the same major influences on their performance (whether those influences be assets or liabilities).

While many players in the Steroid Era have been suspected of using steroids, it is doubtful that every player in Major League Baseball used steroids during that era. Hence, it could be reasonably argued that while certain historical parameters and anomalies – like the use of a "dead" baseball or a 154-game schedule – in previous eras affected *all* players in baseball *equally*, steroid use during the Steroid Era likely affected player performance and statistics *unequally*, since not all players used steroids in that era. As such, it might be argued that the use of z-scores won't control for steroid use as thoroughly as it would some of the other influences in baseball history that affected player performance equally. This is a valid argument, and we do concede that z-scores computed for player statistics in the Steroid Era will not completely control for the influence of steroids. However, we believe that z-scores are still extremely useful in the evaluation of players from the Steroid Era for the following reasons:

- In officially authorized investigations, such as the Mitchell Report (which was recommended by Congress and authorized by Major League Baseball), as well as independent accounts, published in books like *Game of Shadows* by Lance Williams and Mark Fainaru-Wada and *Juiced: Wild Times, Rampant 'Roids, Smash Hits & How Baseball Got Big* by former player Jose Canseco, it has been revealed that steroid use in Major League Baseball was extensive in the Steroid Era. According to Jose Canseco, up to 85% of players in the league may have been using steroids during that period. Even if his estimates are high, it's still

likely that a high percentage of players were using steroids, and thus, a high percentage of players' statistics were affected by steroid use.
- If we use z-scores to evaluate the statistics of players eligible for the Hall of Fame, including those of the Steroid Era who were suspected of using steroids – Barry Bonds, Mark McGwire, Roger Clemens, etc. – the non-unanimous use of steroids in the Steroid Era would actually make it harder for them to achieve z-scores comparable to players of other eras because of how z-scores are calculated:

$$\text{Z-score} = \frac{(Score\ X - Mean)}{Standard\ Deviation}$$

As with any fraction, the larger the denominator, the smaller the overall quotient. In the fraction for z-score computation, the denominator is the standard deviation of the data set from which the individual score (or statistic) comes. As was discussed in Chapter 2, the formula we used for the standard deviation is the following:

$$\text{Standard Deviation}_{[Sample]} = \sqrt{\frac{\Sigma\ (Score\ X - Mean)^2}{N-1}}$$

The greater dispersion of scores in a data set (i.e., the wider the range of scores in a data set), the larger the standard deviation (i.e., the denominator of the quotient used to calculate a z-score), and thus, the smaller the z-score. During individual seasons within the Steroid Era (e.g., the 2001 season in which Barry Bonds hit 73 home runs), if some players were using steroids and some players weren't, this would likely lead to a greater dispersion of scores for all statistics and thus larger standard deviations for all statistics. As a result, the z-scores for each of those statistics would be smaller because the standard deviations would be larger. Had Barry Bonds hit his 73 home runs in an era in which either no one was using steroids, OR, everyone was using steroids, the standard deviation for home runs would likely be much smaller because there would likely be less dispersion of scores around the mean, as players would be closer in ability. In that case, his z-score would be much bigger. So, Barry Bonds and other players of the Steroid Era are actually penalized, not advantaged, regarding their z-scores, because not *all* players were using steroids in that era. Thus, z-scores can be very useful for evaluating the performance of players in the Steroid Era, even if we assume that not all players were using steroids.

How can we use z-scores to evaluate a player's worthiness for the Hall of Fame? This is a tricky question, and there are a lot of potentially valid answers. At present, for players who don't reach one of the golden benchmarks (i.e., 500 home runs, 300 wins, 3,000 hits), there are all sorts of arbitrary criteria, and some sportswriters and media figures have suggested that a player needs to have produced about 10 *Hall-of-Fame-caliber seasons* throughout his career. But what constitutes a *Hall-of-Fame-caliber season*? There are many ways to answer that question. It's controversial. Right now, a player can be voted into the Hall of Fame by the BBWAA with only 75% of the vote, which means as many as 25% of the writers may disagree on whether someone *elected* to the Hall of Fame really deserves to be there. There is no unanimity of thought about what constitutes worthiness for the Hall of Fame, so using z-scores won't necessarily address this subjective issue directly. But here are a few ways it could help:

- The Easy Solution (a.k.a, the Mensa Solution): Identify what it means to be "great" in some other context of society and use that standard (converted into a z-score) as the standard for the Hall of Fame. For instance, for entry into Mensa, the society for people with high IQs, individuals must have an IQ of at least 130,[68] which corresponds to a z-score of 2.0. Hence, the standard of a z-score of 2.0 could be used as one

---

[68] http://www.mensa.org/about-us

identifying a Hall-of-Fame-caliber performance. Perhaps an individual could be considered "worthy" of the Hall of Fame if he had at least 10 seasons in which his z-score for at least one of the most important statistical categories was above a 2.0.

- The Difficult Solution: Compute z-scores for all current Hall of Famers, for all important statistical categories, for each season, and then determine a standard, using z-scores, that at least 90% of existing Hall of Famers meet. For example, if it is determined that 90% of existing Hall of Famers have at least 10 seasons in which they have at least two z-scores of 1.50, and in five of those seasons their highest z-score exceeded 2.50, then use this as the standard for all future candidates for the Hall of Fame.

Other Suggestions for the Steroid Problem

Eradicating baseball of steroid use entirely may not be possible. However, we believe that there are still ways that the sport can be *cleaned up*. One solution that has been suggested would be for the owners and the MLBPA to agree, in their next collective bargaining agreement, to a provision in which a positive test for a banned substance immediately invalidates their most recently signed contract, with players being forced to repay a percentage of their earnings back to their teams. This solution, while it is harsh, punitive, and likely to be successful in drastically reducing steroid use throughout the league, is probably never going to happen. Why? Because the Major League Baseball Players Association, which is one of the strongest unions in the world, would likely never agree to such a provision in a collective bargaining agreement.

In lieu of this proposal being enacted, which is about as likely as a long-term peace agreement in the Middle East, we believe there is another way to drastically reduce steroid use in baseball. Based on the testimony of countless players – both past and present

– who suggest that most of the players who use steroids are only doing so to *keep up with* other players who are also doing it, we believe that there is a genuine desire among many, if not most players, to have a clean game. As such, we believe that if the social pressure to NOT use steroids reaches a tipping point, this goal can be achieved. But, how to reach this tipping point?

One simple way would be for players who are actually clean to voluntarily submit themselves to random blood samples, taken by an independent agency. While this is similar to current drug testing policies, the main difference would be that instead of their blood being tested immediately, their blood would not be tested until years later, when valid tests are available for ALL of the substances used in each era to enhance performance. At present, one of the main drawbacks of current testing procedures is that there is a lag of several years between the availability of a new PED, and the development of a valid, minimally-invasive test for that drug. As such, when players "pass" their drug tests at present, it is still possible that they were using a PED that was so new there weren't any valid tests for it. However, if just a few players – perhaps those who were on track to break any of the vaunted historical records – voluntarily submitted themselves to this type of long-term testing (to eliminate any future doubt or invalidation of their records), we believe that it would lead to a movement of other players doing so, and at some point, NOT doing so might become increasingly viewed in the court of public opinion as a tacit admission to using these drugs.

CHAPTER 8

FUTURE DIRECTIONS

"I look to the future, because that is where I am going to spend the rest of my life."

- George Burns,
Comedian / Actor

As one could imagine, there are countless other applications for using z-scores in the interpretation of sports statistics. In addition to examining the z-scores of players in other team sports, it could also be interesting to compute z-scores for the performance of athletes in Olympic events, like track and field, swimming, cycling, bobsled, speed skating, etc. Just as there seems to be an epidemic in American team sports regarding the use of steroids, there also seems to be rampant use of steroids in individual sports, many of which are part of the Olympics. In addition, advances in equipment and training methods have also made a significant difference in expanding the absolute limits of human accomplishment in athletics. However, using z-scores would help us to evaluate the *relative* performance of modern athletes against their peers and their predecessors: both, in the same sport, as well in relation to other sports. If we were to do this, perhaps we would be better able to answer the question: Who is the greatest athlete of all time?

Outside of sports, using z-scores could be tremendously beneficial in many other areas of life. Adding z-scores to school transcripts could help colleges to evaluate whether the A that a student received in one class is equivalent to the A that another student received in another class, or another school. Wherever you look, there is a relevant application for z-scores. We hope that you will find your own in the personal details of your own life. If enough people can learn to apply this concept – to their own lives and to the things that are important to them – perhaps it could transform the way we think about performance in all areas of life. This transformation can help people to achieve a more sophisticated way of thinking about life in general, which we believe can only be a benefit to a society such as ours.

Z-scores and the SAT Exam

If there is a superordinate goal we have for this book, it is to highlight the need for context when making evaluations and comparisons, as context helps us to understand that everything we seek to examine in life occurs within a milieu of relativism. An allusion to relativism is made by Jesus in the Gospel according to Luke (12:48) where it is written: "For unto whomsoever much is given, of him shall be much required." This quote from the Bible is often used by Christian clergy to explain that God's expectations of each person – in terms of behavior, charity and faith – is *relative* to the resources (material and spiritual) given to that person. The point here is not to spur a religious revival, but rather to highlight that relativism of expectations is a concept that is at least as old as the Bible, and we believe that using relativism as a guiding principle can help us to make fairer assessments and comparisons in many aspects of life: one of which involving the SAT exam.

On the surface, the SAT exam is designed to be a fair and objective assessment of the knowledge acquired in school that is most predictive of college readiness and success. All high school students in the United States taking the exam in a given year take roughly the same exam, and as such, the standardized statistics (e.g., percentiles) reported on SAT transcripts are derived from the mean and standard deviation of the national cohort taking that same exam. Hence, if two individuals from very different parts of the country get the same score on the exam, they would achieve the same percentile (i.e., two students with a combined score of 1000 would each have the same percentile of 50)[69].

While this paradigm seems fair and objective in theory – identical scores yielding identical percentiles, regardless of who achieved those scores and where they live – we believe that the

---

[69] This is an estimate of the percentile rank, as various modifications of the SAT exam have led to combined scores that are slightly higher or lower than 1000 yielding a percentile of 50 over the years.

College Board's attempt to standardize scores in this way leads to a gross mis-measurement of the construct (college readiness) that the test is supposed to assess. Why? Because there is no national school district, with identical standards and resources throughout the country, within which all students taking the SAT exam are educated equally. Hence, it cannot be true that all students taking the exam have an equal chance to achieve the same score on the exam.

Despite there being thousands of school districts across the United States with vastly different levels of economic support and standards of education, high school students across the country all have their SAT scores converted to percentiles using the same *national* set of norms (i.e., mean and standard deviation). Stated more simply, all test takers are compared to the same mean, even though attending school in an underprivileged school district makes it much more difficult for students in that district to achieve a mean score.

On Long Island, NY, the region with which we are both most familiar, the most recent data we were able to obtain[70] show that per pupil instructional spending (PPIS) (i.e., spending on teacher salaries and classroom-related expenses) ranged from $10,274 (in the Floral Park / Bellerose district) to $88,588 (in the Fire Island district) per year. The disparity in this particular statistic highlights just one of the inequities facing students across the country: some kids attend schools in districts with much greater resources for instruction than others.

In addition to differences in PPIS, other factors, like median household income, are also important to examine. Although there is not always a robust correlation between PPIS and SAT performance (as some schools spend their money more or less wisely than others), a family's household income sets the parameters for potential remediation of any obstacles hindering learning and academic

---

[70] 2011 US Census Bureau:
http://factfinder2.census.gov/faces/nav/jsf/pages/index.xhtml

performance. Essentially, the higher a family's household income, the more potential money they have to spend on things like: tutoring, assessment and remediation of learning disabilities, SAT prep courses, academic enrichment activities (e.g., math camp), supplemental learning materials (e.g., laptops, tablets, software, etc.), and many other factors.

Once again, on Long Island, there is a great disparity in median household income across school districts, ranging from $15,863 (Floral Park / Bellerose) to $144,494 (Half Hollow Hills). When considering both PPIS and median household income, it is hard to argue that kids who grow up in the Floral Park / Bellerose school district (PPIS=$10,274; median household income=$15,863) have the same chance of getting an average score on the SAT exam as kids who grow up in the school districts of Fire Island (PPIS=$88,588; median household income=$116,375) or Half Hollow Hills (PPIS=$13,650; median household income=$144,949).

However, the disparity in factors relating to student performance on the SAT exam is relatable to the disparity in factors making it easier to hit a home run in 2001, relative to 1919. Regarding home run totals, our method of leveling the playing field was to compute z-scores for each player using the mean and standard deviation from the player's closest peers: those who played in the same league and same season as the player of interest. Similarly, we believe that converting SAT scores to z-scores for each student using the mean and standard deviation of each student's closest peers – those students who were educated in the same school district could be a more equitable way of assessing students in the college admissions process.

Consider the hypothetical example of "Max" and "Anthony"[71] – two students who obtained the same SAT score of 1150. Max attends school in a district with high PPIS and median

---

[71] The data from this section were obtained from the College Board for the 2011-2012 academic year for two school districts that did not want to be identified.

household income, where the mean SAT score is 1228 (SD=98); Anthony attends school in a district with low PPIS and median household income, where the mean SAT score is 941 (SD=98). For the year in which Max and Anthony take the SAT exam, the mean for the national cohort was 1011 and the standard deviation for the national cohort was 98. Table 8.1 below identifies the z-scores for Max's and Anthony's SAT performance, first when using the mean and standard deviation from the national cohort to compute z-scores, and then when using the mean and standard deviation from Max's and Anthony's respective school districts.

TABLE 8.1: Z-SCORES FOR "MAX" AND "ANTHONY" DERIVED BOTH FROM NORMS OF THE NATIONAL COHORT AND THEIR RESPECTIVE SCHOOL DISTRICTS

STUDENT	SAT SCORE	MEAN & SD OF NATIONAL COHORT	Z-SCORE DERIVED FROM MEAN & SD OF NATIONAL COHORT	MEAN & SD OF RESPECTIVE SCHOOL DISTRICT	Z-SCORE DERIVED FROM MEAN & SD OF RESPECTIVE SCHOOL DISTRICT
Max	1150	Mean = 1011 SD = 98	1.42	Mean = 1228 SD = 98	-0.80
Anthony	1150	Mean = 1011 SD = 98	1.42	Mean = 941 SD = 98	2.13

As shown in Table 8.1, Max and Anthony each achieved the same SAT score, yielding identical z-scores (Z=1.42) when the mean and standard deviation from the national cohort was used to compute them. However, their z-scores were very different when calculated using the mean and standard deviation of their respective school districts. Max had a z-score of -0.80, which was below the mean of the other students in *his* school district, while Anthony had a z-score of 2.13, well above the mean of the other students in *his* school district. Again, relative to the national cohort, their z-scores were identical. However, relative to their closest peers – the other students in their own school district who attended schools with the same resources, facing the same obstacles – Anthony significantly outperformed Max (and the overwhelming majority of students in his school district).

Of course, instead of calculating z-scores, we could have calculated percentiles, as this is a more traditional statistic reported on SAT transcripts; however, we used z-scores for the current analysis for two reasons: a) to be consistent with the data we report in the rest of the book; and b) we believe that for this particular measure, z-scores are easier for the average lay person to understand for the reasons described in Chapter 1. The real innovation in our analyses, which we are proposing as a future area of exploration, is using the mean and standard deviation of school districts to compute standardized statistics (either percentiles, or z-scores) for the SAT exam (or any other national academic achievement test), rather than the mean and standard deviation of the national cohort.

Perhaps, simply replacing the existing method (using the mean and standard deviation of the national cohort) with our proposed method (using the mean and standard deviation of each respective school district) is too radical a change. Perhaps this switch would overcorrect for an existing inequity by creating an inverse inequity. If either of these turn out to be true, we see no harm in the College Board reporting both forms of standardized statistics on SAT transcripts, as more information is likely better than less information when it comes to college admissions boards making decisions on accepting students to their schools.

Gulliver in Lilliput & Brobdingnag

In the end, we believe that the analyses and arguments presented throughout the book are important to consider, not only because they help us to evaluate athletic performance more fairly, but because they speak to the need for more accurate assessments in all areas of life. From a psychological perspective, the markers we use to evaluate our own performance are almost always relative, reflecting the norms and peers that are most salient in our lives. Our conscious recognition of these norms and peers is what sets our expectations, thus contributing to our self-image, self-esteem, and happiness.

For example, in my (John's) private psychology practice I have one patient, whom we will call "Omar," who is dependent on social services and makes less than $20,000 per year. While this level of poverty would lead most people to wake up inconsolably depressed each day, Omar is one of the most optimistic and appreciative individuals I know. Why? Because most of his closest peers – his siblings and friends from childhood – have far less than he does: they are either dead, incarcerated or struggle with a standard of living much lower than his.

In contrast to Omar I have another patient, an adolescent we will call "Lena," whose family has assets in excess of $5 million. Lena, however, lives in an upper-crust neighborhood where her family is at the lower end of the income scale. Though the affluence of Lena's family allows her to enjoy possessions and experiences that less than 1% of her peers across the world can share, she feels like a pauper. Why? Because she does not compare herself to the rest of the world: this is too abstract an exercise for her, as it would be for most of *us*. Rather, in *her* world, among *her* peers, Lena has to make do without many of the things that her closest peers enjoy. For instance, her $350 Coach pocketbook – an exorbitant luxury for most people living outside of her upscale hamlet – is regarded as an inferior, low-budget substitute by her teenage friends who strut around school with $1,000 Louis Vuitton bags. This leads her to feel that she and her family are inadequate – imposters within their community – which contributes to her having a poor self-image and low self-esteem.

The examples of Omar and Lena touch us all because at one time or another we have shared in the perspective of their circumstances. This perspective is analogous to that of Gulliver (from *Gulliver's Travels*), who, despite remaining the same size throughout the book, first feels like a behemoth among the tiny folk of Lilliput and then like a gnome among the giants of Brobdingnag. The analyses and arguments presented throughout the book – about baseball statistics, market power, and SAT performance – strive to make the experiences of Omar and Lena accessible using the familiar pastime of baseball. While we hope that you are now able to appreciate why z-scores offer a superior method for comparing baseball performance throughout history, we would be even happier if the larger lessons of the book – about the importance of context and relativism when evaluating ourselves and others – help you to lead a more peaceful and compassionate existence.

Appendix A

Glossary of Abbreviations

BALCO – Bay Area Laboratory Co-operative

BBWAA – Baseball Writers' Association of America

DH – Designated Hitter

ERA – Earned Run Average

MLB – Major League Baseball

MLBPA – Major League Baseball Players Association

OPS – On-base Percentage (divided by) Slugging

PEDs – Performance Enhancing Drugs

RBI – Runs Batted In

SABR – Society for American Baseball Research

SD – Standard Deviation

WHIP – Walks (plus) Hits (divided by) Innings Pitched

APPENDIX B

REFERENCES

ARTICLES & BOOKS:

Anonymous (2013). "Timeline of Baseball's Steroid Scandal." (http://nbcsports.msnbc.com/id/22247395/)

Author Unknown (1963). "Evolution of the Ball". Rawlings Trade Digest (www.Books.Google.com). Retrieved 26 October 2010.

Canseco, Jose (2006). *Juiced: Wild Times, Rampant 'Roids, Smash Hits & How Baseball Got Big*. It Books; New York.

Creamer, Robert W. (1992), *Babe: The Legend Comes to Life, (2nd Ed.)*. Simon & Schuster; New York.

Dilbeck, Steve (July 23rd, 2013). "Skip Schumaker says Ryan Braun should be suspended for life." *Los Angeles Times*.

Egelko, Bob (February 14, 2007). "Attorney pleads guilty to leaking BALCO testimony". The San Francisco Chronicle.

Enders, Eric (2003). *100 Years of the World Series*. Barnes & Noble Publishing, Inc.;
New York.

Fainaru-Wada, Mark & Williams, Lance. (2007). *Game of Shadows*. Gotham Publishing.

Fainaru-Wada, Mark & Williams, Lance. (2004-12-03). "What Bonds told BALCO grand jury". *San Francisco Chronicle* (Hearst Communications Inc.). Archived from the original on 18 November 2007. Retrieved 2007-10-10.

Gillette, Gary & Palmer, Pete (2007). *The ESPN Baseball Encyclopedia (Fourth Ed.)*. Sterling Publishing Co.; New York.

Leggett, William (1969). "From Mountain to Molehill." *Sports Illustrated: The Vault*. (http://sportsillustrated.cnn.com/vault/article/magazine/MAG108221 1/2/index.htm)

Leventhal, Josh (2000). *Take Me Out to the Ballpark*. Black Dog & Leventhal Publishers, Inc.;
New York.

Levitt, Dan (2005). "Early ERA Titles." SABR's *The National Pastime Research Journal*, 25. (http://research.sabr.org/journals/files/SABR-National_Pastime-25.pdf)

Lockhart, Robert S. (1998). *Introduction to Statistics and Data Analysis for the Behavioral Sciences*. W.H. Freeman, New York.

Major League Baseball. (2012). "MLB's 2012 Regular Season Attendance Ranks as Fifth Highest Ever." MLB.com. (http://mlb.mlb.com/news/article.jsp?ymd=20121004&content_id=3 9483214&vkey=pr_mlb&c_id=mlb)

Naismith, James (1996). *Basketball: Its Origin and Development*, Bison Books.

National Health and Nutrition Examination Survey of 1976-1980 (NHANES II).

Pincus, Arthur, Rosner, David, Hochberg, Len & Malcolm, Chris (2010). *The Official Illustrated NHL History: The Official Story of the Coolest Game on Earth*, Carlton Books.

Sipple, George (July 23rd, 2013). Scherzer: Void Braun's $113M Contract." *The USA Today*.

Schmidt Michael. (June 16, 2009). "Sosa is Said to Have Tested Positive in 2003". *The New York Times.*

Schwartz, Nick (October 24, 2013). "The average career earnings of athletes across America's major sports will shock you." *The USA Today.*

Solomon, Burt (2001). *The Baseball Timeline.* DK Publishing, Inc.; New York.

Strauss, Esther, Sherman, Elisabeth M.S., Spreen, Otfried (2006). A Compendium of Neuropsychological Tests: Administration, Norms, and Commentary: Third Ed. Oxford University Press.

Vaccaro, Frank (2011). "Origins of the Pitching Rotation." *SABR's Baseball Research Journal*, 40, 2. (http://sabr.org/research/origins-pitching-rotation).

WEBSITES:

ESPN No-Hitter History:
http://espn.go.com/mlb/history/nohitters

Baseball Leaderboard Glossary:
www.Baseball-reference.com. Retrieved 2012-05-26:

Mensa:
http://www.mensa.org/about-us

MLB Official Rules:
http://mlb.mlb.com/mlb/official_info/official_rules/objectives_1.jsp

National Center for Health Statistics:
http://www.cdc.gov/nchs/nhanes.htm

National Center for Education Statistics:
http://nces.ed.gov/

Pro-Baseball Statistics:
http://www.mlb.com

Pro-Basketball Statistics:
http://www.basketball-reference.com

Pro-Football Statistics:
http://www.pro-football-reference.com

Pro-Hockey Statistics:
http://www.hockey-reference.com

Sportrac:
www.sportrac.com

2011 United States Census:
http://factfinder2.census.gov

APPENDIX C

FORMULAS

> Key:
> N = Number of scores in a distribution
> X = Score of interest
> Σ = Summation of scores in a given set or distribution

$$\text{Z-score} = \frac{(\text{Score } X - \text{Mean})}{\text{Standard Deviation}}$$

$$\text{Mean} = \frac{(\text{Sum of All Scores})}{N}$$

$$\text{Standard Deviation}_{(\text{Population})} = \sqrt{\frac{\Sigma\,(\text{Score } X - \text{Mean})^2}{N}}$$

$$\text{Standard Deviation}_{(\text{Sample})} = \sqrt{\frac{\Sigma\,(\text{Score } X - \text{Mean})^2}{N-1}}$$

Market Power = (Median Household Income) x (Population)

APPENDIX D

COMPUTATION EXAMPLES

COMPUTATION OF THE STANDARD DEVIATION:
EXAMPLE - CLASS A AND CLASS B (CHAPTER 1, SEE PG. 17)

$$\text{Standard Deviation (SD)} = \sqrt{\frac{\Sigma\,(Score\ X - Mean)^2}{N}}$$

Σ = Sum
N = Number of Scores or Observations

Class A: Computation of the Standard Deviation

Raw Test Score	Mean	(Score X- Mean)	(Score X- Mean)²
100	60	(100-60) = 40	(40x40) = 1,600
100	60	(100-60) = 40	(40x40) = 1,600
100	60	(100-60) = 40	(40x40) = 1,600
100	60	(100-60) = 40	(40x40) = 1,600
100	60	(100-60) = 40	(40x40) = 1,600
20	60	(20-60) = - 40	(-40x-40) = 1,600
20	60	(20-60) = - 40	(-40x-40) = 1,600
20	60	(20-60) = - 40	(-40x-40) = 1,600
20	60	(20-60) = - 40	(-40x-40) = 1,600
20	60	(20-60) = - 40	(-40x-40) = 1,600
Sum of Squares: Σ (Score X- Mean)²			16,000

$$\text{Standard Deviation (SD)} = \sqrt{\frac{16,000}{10}}$$

$$\text{Standard Deviation (SD)} = \sqrt{1,600}$$

Standard Deviation (SD) = 40

Class B: Computation of the Standard Deviation

Raw Test Score	Mean	(Score X- Mean)	(Score X- Mean)^2
70	60	(70-60) = 10	(10x10) = 100
70	60	(70-60) = 10	(10x10) = 100
70	60	(70-60) = 10	(10x10) = 100
70	60	(70-60) = 10	(10x10) = 100
70	60	(70-60) = 10	(10x10) = 100
50	60	(50-60) = - 10	(-10x-10) = 100
50	60	(50-60) = - 10	(-10x-10) = 100
50	60	(50-60) = - 10	(-10x-10) = 100
50	60	(50-60) = - 10	(-10x-10) = 100
50	60	(50-60) = - 10	(-10x-10) = 100
Sum of Squares: Σ (Score X- Mean)^2			1,000

Standard Deviation (SD) = $\sqrt{\dfrac{1,000}{10}}$

Standard Deviation (SD) = $\sqrt{100}$

Standard Deviation (SD) = 10

INDEX

Aaron, Hank, xi, 113, 116, 152
Aron, Dr. Art, vii
Austin Powers, 3
Baker, Frank, 49, 87
BALCO, 70, 179
Bando, Sal, 113, 120
Baseball Writers' Association of America (BBWAA), xi, 116, 117, 118, 119, 120, 122, 123, 124, 125, 126, 127, 128, 130, 131, 162, 165, 177
Bautista, Jose, xi, 78, 85, 99
Beane, Billy, 8, 9, 13, 133, 134, 135, 136, 138, 139, 143, 150
Bedrosian, Steve, 65
bell curve, 20, 22
Big Lebowski, The, 80
Biogenesis, 76
Blue, Vida, 113, 120
Bol, Manute, 6
Bonds, Barry, ix, xi, 4, 5, 10, 69, 70, 71, 73, 84, 85, 89, 99, 101, 102, 103, 104, 138, 152, 157, 158, 164, 165, 179
Bonham, Tiny, 55, 97
Braun, Ryan, 76, 77, 179, 180
Brees, Drew, 154
Brett, George, 66, 83, 99, 101
Brodeur, Martin, 156
Brown, Mordecai, 91
Bryant, Kobe, 144
Burns, George, 168
Cabrera, Miguel, 76, 78, 87, 89, 99
Canseco, Jose, 68, 70, 163, 179
Carew, Rod, 60, 82, 83, 99
Carlton, Steve, 61, 93, 180
Carty, Rico, 60, 83
Cash, Norm, 60
Chamberlain, Wilt, 155
Chandler, Spud, 55, 91

Chesbro, Jack, 50, 93, 100
Chronic Traumatic Encephalopathy (CTE), 146
Clemens, Roger, xii, 75, 113, 124, 164
Conte, Victor, 71
Cravath, Gavvy, 33, 49, 85, 87, 99, 101, 157
Darvish, Yu, 79, 95
Davids, L. Robert, 8
Davis, Al, 92
Davis, Mark, 65
Davis, Tommy, 60, 87
Dean, Dizzy, 55, 93
Dickerson, Eric, 154
DiMaggio, Joe, 51, 52, 152
Dowling College, v, 193
Drysdale, Don, 58
Duffy, Hugh, 45, 82, 83, 89, 108, 109
Eckersly, Dennis, 65, 113, 114, 125
Einstein, Albert, 6
Ellerman, Troy, 71
Elliot, Brian, 156
Ellsbury, Jacoby, 113, 114, 126
Epstein, Theo, 8
Fainaru-Wada, Mark, 70, 71, 163, 179
Feller, Bob, 52, 55, 95, 100
Field of Dreams, 40
Fingers, Rollie, 65, 113, 114, 121, 122
Flood, Curt, 62
Foreman, George, 6
Foster, George, 66, 87
Foxx, Jimmie, 52
Freehan, Bill, 113, 119
Gagne, Eric, 65, 75
Game of Shadows, 70, 71, 163, 179
Garciaparra, Nomar, 73, 83
Gehrig, Lou, 52, 54, 87
Giambi, Jason, 129, 130
Gibson, Bob, 41, 57, 59, 61, 90, 91, 97, 107, 113, 118

Glavine, Tom, 84
Gooden, Dwight "Doc", 67, 91, 93
Gossage, Rich "Goose", 63
Graham, Moonlight, 3
Greenberg, Hank, 52
Gretzky, Wayne, 5, 151, 156, 157, 158, 159
Groat, Dick, 113, 117
Guidry, Ron, 67, 91, 93, 94, 97
Gwynn, Tony, 73, 83, 99
Haas Jr., Walter A., 133
Hacker, Warren, 55, 97
Hainsworth, George, 156
Hall of Fame, ix, xi, xiii, 11, 51, 71, 82, 105, 106, 108, 109, 131, 159, 160, 161, 162, 164, 165, 166
Hayes, Willie Mays, 148
Henderson, Ricky, 113, 114, 121, 122
Hernandez, Felix, 77, 79, 91, 97
Hernandez, Willie, 65, 113, 114, 123, 124
Hershiser, Orel, 152
Hofmann, Ken, 133
Holyfield, Evander, 6
Hornsby, Rogers, 54, 83, 89, 99
Howard, Ryan, 78, 85, 87
Hrbek, Kent, 113, 123
Hubbell, Carl, 55, 91
Hunter, Catfish, 93
IQ, 6, 21, 29, 30, 31, 104, 138, 158, 165
James, Bill, 8, 9, 55, 60, 87, 93, 134, 135, 180
Johnson, Calvin, 154
Johnson, Randy, 74
Johnson, Walter, 50, 93, 95, 97, 100, 101
Jones, Chipper, 78, 83
Kanehl, Roderick "Hot Rod", 57
Keefe, Tim, 45, 50, 91, 97
Kershaw, Clayton, 77, 79, 91, 97, 113, 127
Kingman, Dave, 66, 85
Klein, Chuck, 52

Korver, Kyle, 155
Koufax, Sandy, 57, 61, 95, 100, 113, 117
Lajoie, Nap, 49, 83, 99, 101, 108, 109
Lane, Night Train, 154
LaRussa, Tony, 63
Lennon, David, ix
Leonard, Dutch, 50, 91, 100
Lewis, Michael, 8, 13, 133, 134
Lincecum, Tim, 79, 95
Lyle, Sparky, 64
MacPhail, Lee, 62
Maddux, Greg, 74, 84, 91, 97, 100
Mann, Terence, 40
Manning, Peyton, 154
Marichal, Juan, 57, 58
Maris, Roger, xi, 4, 5, 60, 69, 84, 85, 87, 102, 104, 152
Market power, 140, 183
Marshall, Mike, 64
Martinez, Edgar, 71
Martinez, Pedro, ix, 5, 74, 91, 96, 97, 100, 101, 105, 128, 129, 130
Matthewson, Christy, 93, 95, 97, 100, 101
Mattingly, Don, xii, 66, 87, 113, 124
Mauer, Joe, 78, 83
Mays, Willie, 51, 57, 60, 85
Mayweather Jr., Floyd, 6
McCovey, Willie, 60, 89
McGee, Willie, 66, 83
McGwire, Mark, xi, 4, 68, 69, 70, 71, 84, 164
McLain, Denny, 61, 90, 92, 93, 100, 107, 113, 119
McNally, Dave, 61, 62, 97
mean, 15, 16, 17, 20, 21, 22, 23, 24, 25, 26, 28, 29, 30, 82, 88, 90, 104, 108, 109, 117, 118, 121, 122, 137, 138, 143, 148, 154, 165, 170, 171, 172, 173, 174
Mensa, 30, 138, 165, 182
Messersmith, Andy, 62
Mitchell Report, The, 75, 76, 163
Mitchell, George, 75

MLBPA, 62, 68, 76, 77, 177
Moneyball, 8, 9, 13, 132, 133, 134, 149
Moss, Randy, 154, 157, 159
Murphy, Dale, 4
Musial, Stan, 51, 52
Nevid, Dr. Jeffrey, vii
Newcombe, Don, 113, 116
Ortiz, David, 71
Ott, Mel, 52
Ovechkin, Alex, 144, 145
Palmer, Jim, 58
Park, Chan Ho, 4
percentile, 23, 24, 27, 28, 29, 30, 170, 171, 174
Pesky, Johnny, 52
Pettitte, Andy, 75
Pucket, Kirby, 113, 125
Pujols, Albert, 78, 89
Radbourn, Charley, 45, 50, 93, 95, 106, 107, 108, 109
Ramirez, Manny, 73, 87
Rice, Jim, 66, 85
Richard, J.R., 67, 95, 100, 101
Ripken Jr., Cal, 152
Robison, Jackie, 52
Rodgers, Aaron, 144, 145
Rodriguez, Alex, 73, 76, 85, 144, 145
Rodriguez, Ivan, 128, 129, 130
Rose, Pete, 113, 118, 152
Ruth, Babe, ix, xi, 5, 33, 47, 49, 51, 52, 54, 84, 85, 89, 99, 101, 102, 103, 104, 106, 153, 157, 158, 159
Ryan, Nolan, 61, 67, 95, 100, 101, 152
sabermetrics, 8, 88, 150
SABR, 8, 9, 134, 135, 180, 181
SAT exam, 10, 23, 24, 170, 171, 172, 173, 174
Schaub, Jordan, 133
Schell, Michael, 7, 8, 9
Scherzer, Max, 77, 180
Schmidt, Mike, 66, 71, 89, 99, 181

Schopenhauer, Arthur, 160
Schott, Stephen, 133
Schumaker, Skip, 77, 179
Seaver, Tom, 57
Selig, Bud, 75
Seymour, Cy, 49, 83
Sheen, Charlie, 92, 148
Sisler, George, 54, 83
Sosa, Sammy, xi, 4, 69, 70, 71, 73, 84, 87, 181
St. John's University, vii, 4, 193
standard deviation, 15, 16, 17, 18, 21, 22, 23, 25, 28, 30, 103, 104, 122, 149, 164, 170, 171, 172, 173, 174
Standard Deviation Formula - Sample, 16, 164
Standard Deviation Formula - Population, 16, 183, 184, 185
Stanton, Giancarlo, 113, 127
Stockton, John, 155, 157
Stony Brook University, vii, 193, 195
Strahan, Michael, 154, 157, 159
Sutter, Bruce, 65
Tejada, Miguel, 75
Thomas, Frank, 73, 99
Thompson, Sam, 45, 87, 99, 101
Tiant, Luis, 61, 90, 91
Tudor, John, 67, 97
Tyson, Mike, 6
Vance, Dazzy, ix, 55, 95, 100, 101, 104, 105, 157
Vaughn, Rick "Wild Thing", 148
Verlander, Justin, 77, 79, 93, 97, 100, 113, 126
Waddell, Rube, 50, 95, 100, 101
Webb, Brandon, 79
Welch, Bob, 74, 93, 100
Williams, Lance, 70, 71, 163
Williams, Ted, x, 51, 52, 110
Williamson, Ned, 45, 51, 85, 99, 101
Wilson, Hack, 52, 54, 85, 87
Wolff, Lewis, 133
Yi, Dr. Richard, vii

Young, Cy, 152
Young, George, 132
Youth, Dr. Robert, v
Zimmerman, Heinie, 49, 89
Z-score Formula, 15, 183

ABOUT THE AUTHORS

Dr. John G. Cottone is a clinical psychologist in private practice in Stony Brook, NY. He received his Ph.D. in clinical psychology from St. John's University and previously earned a master's degree in biopsychology from Stony Brook University. He is the author of numerous peer-reviewed research publications, including articles and book chapters on topics in psychology, psychiatry and neuroscience. He is the author of a 2013 self-help book entitled *Who Are You? Essential Questions for Hitchhikers on the Road of Truth* (Story Bridge Books) and has had vocational and avocational contributions to *The New York Times* and *The Washington Post.* He is also an avid baseball fan, a member of Red Sox Nation, and was a four-year scholar-athlete at Dowling College, where he pitched on the school's baseball team.

Jason Wirchin earned his B.A. in Political Science, along with a Minor in Journalism, from Stony Brook University in 2010. He is currently a television news producer for News 12 The Bronx, and has also established himself as a digital content producer with the News 12 Networks. He has been published several times by both national and local papers, including *The New York Times*, *New York Daily News* and *Newsday;* and is a frequent caller to WFAN. Jason takes great pride in the history of our national pastime and is a religious fan of the New York Mets. The eldest of three, he grew up in Huntington Station, Long Island and lives with his wife, Ellen, in New York City.

NOTES

Notes

Notes

Check out other great publications from **Story Bridge Books**:

http://www.storybridgebooks.com

Who Are You?
Essential Questions for Hitchhikers on the Road of Truth
by John G. Cottone, PhD © 2013

Who Are You? Essential Questions for Hitchhikers on the Road of Truth explores the questions we need to ask – about the psychology of human behavior, politics, science, metaphysics, and the mysteries of God – to live lives of meaning.
It inspires readers to gently spiral through deeper states of contemplation and self-inquiry using reflective questioning and Socratic dialogues. *Who Are You?* is a meditation companion, a catalyst for group discussion, a personal mirror for honest glimpses at the soul and an invitation for self-growth.

Book of the Sky God
by Laura Markowitz © 2013

The gods are coming back to Earth, and they're not particularly happy. Five unlikely teenagers and one immortal shape-shifter are the only ones who can save humanity. **Ram Rajathani** thinks he knows everything about his best friend, but he doesn't know that Henry has been turned into a part-time zombie, or that Henry's got a crush on a girl he recently met online. The girl, **ComixChik8**, is torturing Henry on purpose, but what she really wants to know is what she's supposed to do with the cursed blue iguana that mysteriously appeared in her school locker. **Henry Lipton** has no idea he's a part-time zombie. He thinks his biggest problem is girl-o-phobia. Ram's sister **Laila** has a dangerous Mayan artifact stashed in the back of her closet. She suspects it can save humanity from the gods, but she has no idea how. Her older sister, **Nina**, isn't afraid of gods, but she's desperate to make some friends. She's about to discover she has mad spy skills. In the meantime, the secretive **Brotherhood of the Prophecy** rescheduled the end of the world, and now they plan to cash in on it. These five teenagers must race against time to outwit the Brotherhood and save humanity from the gods. No biggie.

www.ingramcontent.com/pod-product-compliance
Lightning Source LLC
Chambersburg PA
CBHW051124160426
43195CB00014B/2330